Conversion of U.S. Customary Units to Metric Units

Conversion Factors (listed alphabetically)

To convert from	To	Multiply by*
kip-foot (kip · ft)	newton-meter (N · m)	1356
kip per foot (kip/ft)	kilogram per meter (kg/m)	1488
kip per square inch (kip/in², ksi)	kilogram per square centimeter (kg/cm²)	70.31
lambert [1 lumen/cm²] (L)	candela per square meter (cd/m²)	3183
mil [0.001 in]	centimeter (cm)	0.002540
mile [Inter.] (mi)	kilometer (km)	1.609
mile per hour (mi/h, mph)	kilometer per hour (km/h)	1.609
	meter per second (m/s)	0.4470
ounce [avoirdupois] (oz)	gram (gm)	28.35
pound [avoirdupois] (lb)	kilogram (kg)	0.4536
pound per cubic foot (lb/ft³)	kilogram per cubic meter (kg/m³)	1.602
pound per foot (lb/ft)	kilogram per meter (kg/m)	1.488
pound per square foot (lb/ft²)	kilogram per square meter (kg/m²)	4.882
pound per square inch (lb/in²)	kilogram per square centimeter (kg/cm²)	2.927
rod [16.5 feet]	meter	5.029
quart [U.S. liquid] (qt)	liter (L)	0.9463
square foot (ft²)	square meter (m²)	0.09290
square inch (in²)	square centimeter (cm²)	6.452
square mile [Inter.] (mi²)	square …	2.590
		259.0
square yard (yd²)		8.361
therm [100,000 B…]		1.055 x 10⁸
ton [long = 2240 …		016
[short = 2000 …		07.2
ton [of refrigeration…		517
watthour (Wh)		3600
yard (yd)		0.9144

(Table is continued in back of the book.)

ARCHITECTURAL AND ENGINEERING CALCULATIONS MANUAL

ARCHITECTURAL AND ENGINEERING CALCULATIONS MANUAL

ROBERT BROWN BUTLER

McGraw-Hill Book Company

New York St. Louis San Francisco Auckland Bogotá Hamburg
London Madrid Mexico Montreal New Delhi
Panama Paris São Paulo Singapore Sydney Tokyo Toronto

Library of Congress Cataloging in Publication Data

Butler, Robert Brown
    Architectural and engineering calculations manual.
    Includes index.
    1. Engineering mathematics.   2. Structural
engineering—Mathematics.   I. Title.
TA330.B87 1984   721'.0212   83-7934
ISBN 0-07-009363-6

    67890   KGP/KGP   8987

ISBN 0-07-009363-6

The editors for this book were Joan Zseleczky and Irene Curran, the design supervisor was Mark E. Safran, and the production supervisor was Thomas G. Kowal-
czyk. It was set in Times Roman by University Graphics, Inc.

Printed and bound by The Kingsport Press. Layout by Small Kaps Associates,
Inc.

**To Janis Youngstrum Butler,
my wife and friend**

CONTENTS

PREFACE

This volume is a pocket-sized collection of mathematical problems that occur during the design of commercial and residential architecture. Each calculation deals with a circumstance governing the design of an architectural space, structural member, plumbing or electrical component, lighting fixture, acoustic assembly, or HVAC detail. Every example describes an actual situation, lists the formula needed to size the architectural part, explains each unknown, and then gives the answer. The problems are easy to find, exist independently of each other, and do not require outside reading to solve. The book itself is designed to slip into a glove compartment, a purse, and even a coat pocket and can be easily consulted during a conversation or lecture.

This book can serve many people. It would aid an architect who needs to find comfortable dimensions for an unusual space or wants to size a complicated structural component without having to spend hours scanning pages of antiquated texts and relearning subjects forgotten decades ago. It would help an engineer who, upon receiving a soil boring report, finds a stratum of softer soil under the harder ground upon which a footing will rest and now must quickly design a larger footing of proper size. It would guide a consultant in illumination who needs to find the coefficient of utilization of an interior space but who long ago lost any affinity for wading through the tedious calculations this work traditionally requires. A specialist in heating design would find his or her calculations similarly reduced. For apprentices and students, this reference provides instant expertise in all the above professional tasks and many more. The book would also advise a realtor or prospective homeowner who wants to know if a driveway can be built on a piece of hilly property outlined on a topographic map. It would help a government administrator who wants to know the ideal square footage for a new elementary school. It would be a blessing to a professor who, while instructing his pupils, suddenly needs to deal with earthquake stresses—taking his "AEC" from nearby, he could nimbly locate the information without a hitch in his discourse. And it would be a godsend to a homeowner who wants to build a new porch behind the house or install solar panels on the roof. Indeed, this manual should quickly return its purchase price in saved time and reduced anxiety to anyone who desires to command the mathematics of architecture.

Using this volume is simple. If you have a problem, look it up in the index, turn to the page and pull out your calculator. Read the problem in the book, and by analogy relate your quantities and unknowns to the ones in the equation. Make sure your quantities are in the same units as those in the equation. Then press the buttons to find the answer. The whole process is designed so that if a colleague telephones you to discuss an architectureal problem, while the two of you are on the phone fleshing out the details, you can be looking up the proper formula and plugging in the unknowns. By the time the discussion has ended, you will have the answer. A few of the problems are more complicated and require two or more steps to solve.

It has been said that creativity is two precent inspiration and ninety-eight percent perspiration. This book, by reducing the amount of perspiration you spend on a project, will increase your inspiration to design more creatively.

In any book, the author requires the help of many others to make the publication successful. My foremost acknowledgment is to my wife Janis, who discussed with me every aspect of this volume's contents during each stage of its development. I next thank Joan Zseleczky, the book's acquisitions editor, who offered much advice and persuaded me to perform extra labor that will be to this work's everlasting benefit. I also thank Irene Curran for her fine and persistent editing and Jeremy Robinson for steering me into this literary relationship.

Finally, I thank all the clients, carpenters, contractors, and consultants I have worked with during the last fifteen years who gave me a clue that led to a note that later surfaced as part of a formula, table, or statement in this volume. They are the unsung heroes who give it molecular weight.

Robert Brown Butler

INTRODUCTION

GENERAL

This volume contains mathematical problems that are useful for sizing an architectural component during the design of commercial or residential architecture. Because of the book's miniaturized format, derived terms are built into each equation and derivations of equations are not explained in the text. What appears here is succinct and immediately utilizable.

OBJECTIVE

For small projects, this volume may replace nearly all consultation with other professionals. On large projects, during the preliminary design phase, the examples given in this book may be used to clarify the nature of the architecture quickly and accurately. During design development, this book should enable the architect to comprehend more easily his or her consultants' calculations and thereby expedite all phases of the work. For feasibility projects, its information will serve to define the nature of the work quickly and clearly. The volume may also be helpful to people involved in occupations related to architecture, such as engineers, contractors, subcontractors, lawyers, realtors, students, and candidates for architectural registration.

ADDITIONAL REFERENCES

In order to solve certain intricate or unusual architectural problems, the reader may find use for the additional references listed below.

American Institute of Steel Construction: *AISC Manual,* 8th ed., American Institute of Steel Construction, Chicago, 1980.

American Institute of Timber Construction: *Timber Construction Manual,* 2d ed., Wiley, New York, 1974.

American Society of Heating, Refrigeration, and Air Conditioning Engineers: *ASHRAE Handbook of Fundamentals,* Atlanta, 1981.

Callender, John Hancock (ed.): *Time-Saver Standards,* 5th ed., McGraw-Hill, New York, 1974.

Merritt, Frederick S.: *Building Construction Handbook,* 3d ed., McGraw-Hill, New York, 1975.

National Association of Plumbing-Heating-Cooling Contractors and the American Society of Plumbing Engineers: *National Standard Plumbing Code,* McGraw-Hill, New York, 1957.

National Fire Protection Association: *National Electrical Code,* Boston, 1975.

Parker, Harry: *Simplified Design of Reinforced Concrete,* 4th ed., Wiley, New York, 1976.

Various manufacturer's product catalogs, such as open web steel joist tables, mechanical heating equipment specifications, lamp output data, and the like may also be useful.

For mathematics involving weather data, the reader should possess local climatic data. An encyclopedic and inexpensive source of national climatic data is

U.S. Department of Commerce: *Comparative Climatic Data,* National Climatic Center, Federal Building, Asheville, North Carolina 28801. Telephone: (704)258-2850.

THE CALCULATOR

Vital to the use of this manual is the modern pocket calculator, in particular one that displays trigonometric, logarithmic, exponential, and memory functions.

In this book, all calculations are taken to three significant figures.

SYMBOLS

In each equation throughout this book each term is represented by a letter symbol. Familiar quantities employ the usual letters (such as *b* for the width of a beam), but many values are symbolized by the first letter of the most important word that describes them. Thus, one letter may denote different values in different equations.

Many symbols and abbreviations common to mathematics and architecture have been standardized throughout the text. They are listed below.

Throughout this book, the reader must take special care to use the same measuring units as used in each equation. For example, if a quantity is described here in feet and your data is in inches, be sure to convert your data to feet before using the equation.

Mathematical symbols

Symbol	Meaning
$=$	Left side of the equation equals the right side.
$\approx$	Left side of the equation nearly or approximately equals the right side.
$\geq$	Left side of the equation is equal to or greater than the right side.
$\leq$	Left side of the equation is equal to or less than the right side.
$>$	Left side of the equation is greater than the right side.
$<$	Left side of the equation is less than the right side.
$\rightarrow$	Implies, leads to; such as the conclusion involving P leads to ($\rightarrow$) Q.

Symbol	Meaning
$\perp$	Perpendicular to, or normal to; two lines or flat surfaces meet each other at a 90° angle.
$\parallel$	Parallel to; two lines or flat surfaces are parallel to each other.
$\sqrt{A}$	Square root of A; also, $A^{0.5}$.
A^n	A to the nth power, such as $A^{1.875}$, which is A raised to the 1.875th power.
$[A]$	Use only the integer part of A; for example, [2.49] means use only the integer 2: [2.49] = 2.
sin A	Sine of the angle A; in a right triangle, sin A = opposite side/hypotenuse.
cos A	Cosine of the angle A; in a right triangle, cos A = adjacent side/hypotenuse.
tan A	Tangent of the angle A; in a right triangle, tan A = opposite side/adjacent side.
$\sin^{-1} B$	Arcsine B, or sine of the angle whose value is B; if sin $A = B$, then $\sin^{-1} B = A$; the same holds true for $\cos^{-1} B$ and $\tan^{-1} B$.

Abbreviations

A	Ampere, or amperes
BTU	British thermal unit, the amount of heat required to raise the temperature of 1 lb of water 1° Fahrenheit
C	Celsius
cmil	Circular mil, a unit of cross-sectional area measure for electrical wire; the number of circular mils in a one-inch-diameter wire equals 1,000,000 (wire diameter in thousandths of an inch)^2

Abbreviations (Continued)

CU	Coefficient of utilization, the ratio of useful light to actual light in a room
in^3	Cubic inch, or inches
yd^3	Cubic yard, or yards
dB	Decibel, or decibels
dia.	Diameter
F	Fahrenheit
fc	Footcandle, or footcandles
ft	Foot, or feet
ft^2	Square foot, or feet
ft^3	Cubic foot or feet
ft/sec	Foot, or feet, per second
ft/min	Foot, or feet, per minute
ft^2/min	Square foot, or feet, per minute
ft^3/min	Cubic foot, or feet, per minute
ft·lb	Foot pound, or pounds
f/ft	Fall per foot, such as ½-in f/ft, which means ½ in pitch downward for every horizontal foot of pipe outward
FU	Fixture unit, a unit for measuring the relative rate of water flowing into or out of a plumbing fixture
gal	Gallon
gal/min	Gallons per minute
hr	Hour, or hours
Hz	Hertz, or cycles per second; a unit of frequency for alternating current electricity; also, the vibration frequency of a sound, in cycles per second

Abbreviations (Continued)

IIC	Impact isolation class, a unit for measuring the amount of solid-borne sound absorbed by a type of building construction
in	Inch, or inches
in^2	Square inch or inches
in^3	Cubic inch or inches
in·lb	Inch pound or pounds
K	Kelvin; Celsius degrees on an absolute scale; $0°C = 273$ K
kip	1000 pounds
kip·ft	Kip foot or feet
kip·in	Kip inch or inches
kips/lin ft	Kips per linear foot
$kips/ft^2$	Kips per square foot
$kips/in^2$	Kips per square inch
kWh	Kilowatthour, 1000 watts per hour
lb	Pound, or pounds
lb/in^2	Pounds per square inch
lb/ft^2	Pounds per square foot
lb/ft^3	Pounds per cubic foot
lb/lin ft	Pounds per linear foot
lin ft	Linear foot, or feet
log	Logarithm; in this book, all logarithms are to the base 10
lm	Lumen, or lumens
max.	Maximum
min.	Minimum
mi	Mile, or miles
mi^2	Square mile

Abbreviations (Continued)

mi/hr	Miles per hour
min	Minute, or minutes
NG	No good; the value under consideration is not acceptable
nom.	Nominal; a common but slightly excessive method of describing lumber sizes; for example, a 2 x 10 nom. has actual dimensions of 1½ by 9¼ inches
o.c.	On center, or center-to-center; a method of describing dimensions to the center lines of architectural components
OK	Okay; the value under consideration is acceptable
R	Rankine; Fahrenheit degrees on an absolute scale; $0°F$ = approx. $460°R$
sec	Second, or seconds
STC	Sound transmission class; a unit for measuring the amount of airborne sound absorbed by a type of building construction
V	Volt, or volts
W	Watt, or watts
yd	Yard, or yards
yd²	Square yard or yards
yd³	Cubic yard or yards
yr	Year, or years
Δ	Deflection in a structural member
°	Degrees; in this book, fractions of degrees are not expressed as minutes and seconds but are in decimal form; for example, $15°36' = 15.6°$.
#	Number size of a reinforcing rod; for example, #8 rebar is a number 8 reinforcing rod, which has a diameter of 1 inch.

Abbreviations (Continued)

%	Percent.
?	Refers to a quantity that at present is unknown.
+	Indicates the preferred value.

METRICATION

In the near future, the United States may adopt the metric system of measure. Considering this, a list of equivalent measures relating English to metric values is presented on the inside cover.

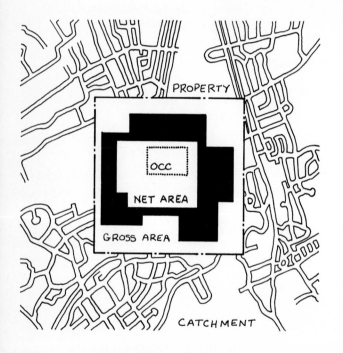

Room in building on property on map.

DESIGN

GENERAL

Architectural design is often considered to be complicated and subjective. Yet many parts of this process can be distilled into mathematical formulas that offer simple, finite analysis. This chapter provides such equations for defining certain aspects of architectural space.

LAND USE

The users of a piece of architecture often come from an easily defined geographical area known as a *catchment*. By knowing a catchment's statistics relevant to a proposed building's use, a designer can more easily estimate the structure's optimum floor area.

Catchments

Example 1 An owner of commercial property plans to build a supermarket. He figures he can attract people living up to 1¼ mi away. You have local data from the U.S. Census Bureau and the U.S. Department of Labor Statistics which says the area contains 4628 people per square mile and that each person spends $1413 per year on food and related items. What is the total market potential for this kind of facility in this area?

$$C = AUI$$

C = catchment potential, ? dollars per year total market potential

A = area of catchment, $A = \pi r^2 = \pi(1.25^2) = 4.91$ mi^2

U = units per area (population per square mile from census data), 4628 people per square mile

I = input per unit (annual food and sundry expenses per person from Department of Labor data), $1413 per year

$$C = 4.91 \times 4628 \times 1413$$
$$= \$32,100,000 \text{ per year}$$

NOTE: To determine the market potential for the facility to be designed, subtract from the total market potential the market penetration rate of existing stores in the catchment area. See "Buildable Floor Area" below.

Example 2   A new suburban development requires an elementary school. The development will have 400 lots, and there are 120 existing residences that will also lie in the new school district. From local census data, 0.8 children per family in the area are of elementary school age. For how many children should the school be designed?

$$C = AUI$$

C = catchment potential, ? students in district
A = area of catchment, $400 + 120 = 520$ lots
U = units per area, 1 family per lot
I = input per unit, 0.8 school age children per family

$$C = 520 \times 1 \times 0.8$$
$$= 416 \text{ students}$$

Buildable Floor Area

Example 1   A businesswoman who plans to build a supermarket knows such facilities generate about $165 in annual sales per square foot of floor area. She has learned that the market potential for this building in its primary market area is $32.1 million per year and that this area already contains 171,000 ft^2 of similar facilities. What is the optimum floor area of the supermarket?

$$B = \frac{C - FU}{U}$$

B = buildable floor area, ? ft^2
C = catchment potential (market potential in dollars per year), $32,100,000
F = floor area of competing facilities, 171,000 ft^2
U = output per unit area (annual sales per square foot), $165 per year

$$B = \frac{32,100,000 - 171,000 \times 165}{165}$$
$$= 23,500 \text{ ft}^2$$

Example 2   The county school board knows a new elementary school facility should be designed for 416 pupils and that its total floor area should be about 65 ft^2 per pupil. What is the optimum floor area of the school?

$$B = \frac{C - FU}{U}$$

B = buildable floor area, ? ft^2
C = catchment potential, 416 students
F = floor area of competing facilities, 0 (no competition within catchment)
U = output per unit area (number of students per square foot), 1/65
  = 0.015

$$B = \frac{416 - 0 \times 0.015}{0.015}$$
$$= 27,700 \text{ ft}^2$$

NOTE: These land use equations may be used to solve problems such as the following:

A cafe is planned for the bottom floor of a building located on the corner of a busy downtown intersection. What is the optimum floor area for the cafe?

A sparsely populated county of 1200 mi² that has no hospital wants
to build one. How many rooms should the hospital have?

A small airline company wants to operate a shuttle service between
a major urban area and a city 100 mi away that has no airport.
Would it be profitable to build a small airport there?

SITEWORK

The equations below concern the manipulation of the environs immediately surrounding a piece of architecture.

Excavation

The earth being excavated to create the basement of a new house is
wanted for surrounding landscaping. If the excavated volume is shaped
as shown in Fig. 2-1, how many cubic yards of earth are available?

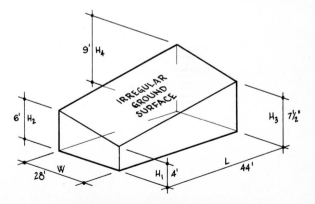

FIGURE 2-1 Excavation with dimensions.

$$V = \frac{LW}{4}(H_1 + H_2 + H_3 + H_4)$$

V = volume of earth, ? ft^3
L = length of volume, 28 ft
W = width of volume, 44 ft
H_1 = height of volume at first corner, 4 ft
H_2 = height of volume at second corner, 6 ft
H_3 = height of volume at third corner, 7.5 ft
H_4 = height of volume at fourth corner, 9 ft

$$V = \frac{28 \times 44}{4}(4 + 6 + 7.5 + 9)$$
$$= 8160 \text{ ft}^3$$

To change to cubic yards: 1 yd^3 = 27 ft^3.

$$V = \frac{8160}{27} = 302 \text{ yd}^3$$

Refer to Fig. 2-2 for excavation formulas.

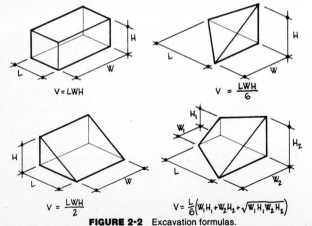

FIGURE 2-2 Excavation formulas.

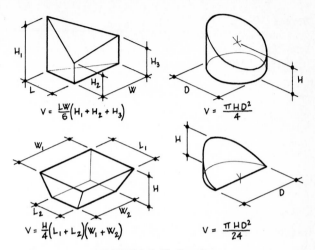

$$V = \frac{LW}{6}\left(H_1 + H_2 + H_3\right)$$

$$V = \frac{\pi H D^2}{4}$$

$$V = \frac{H}{4}\left(L_1 + L_2\right)\left(W_1 + W_2\right)$$

$$V = \frac{\pi H D^2}{24}$$

FIGURE 2-2 *(Continued)*

Complex excavation volumes There are basically two kinds of complex excavation volumes: (1) volumes of *warped or irregular surfaces* and (2) volumes with *complex voids*. With warped or irregular surfaces, it is easiest to average out the irregular surface with a flat plane, as shown in Fig. 2-3. In the case of complex voids, break them down into simple individual volumes and then calculate the size of each.

FIGURE 2-3 Irregular excavation.

Slopes

On a topography map, the hillside contours of a possible homesite are an average of 28 ft apart and the contour elevation interval is 5 ft. Is the slope gentle enough for a driveway?

$$S = 100 \frac{C}{D}$$

S = slope of terrain, ?% grade
C = elevation interval of contours, 5 ft
D = horizontal distance between contours, 28 ft

$$S = 100 \times \frac{5}{28}$$
$$= 17.9\%$$

Is this slope gentle enough for a driveway? Consult Table 2-1.

TABLE 2-1   ALLOWABLE SLOPE GRADES

Slope surface	Grade, %
Appears level to the eye	0–4
Easy grades, suitable for construction	4–10
Steep grades	10+
Grassy recreation areas	1–3
Walks	0–4
Pedestrian ramps	4–8½
Side inclines of curb cutouts	4–16⅔
Parking areas	1–5
Parking access ramps:	
Unroofed, curved	1–6
Unroofed, straight	1–10
Roofed, curved	1–12
Roofed, straight	1–20

TABLE 2-1   ALLOWABLE SLOPE GRADES *(Continued)*

Slope surface	Grade, %
Driveways:	
Dry	0–20
Wet	1–10
Icy	1–4
Roads	0–8
In hilly areas with no winter ice	0–12
Grassy slopes	1–25
Slopes planted with ground cover	0–50
Susceptible to erosion	50+
Ground adjacent to buildings: sloping down away from exterior wall	2–50
Drainage, adequate:	
Small paved areas	0.3+
Unpaved areas	1+
Large paved areas	1+
Drainage ditches	2–10
Storm drains	3–10
Sanitary sewers:	
4-in diameter	2–15
8-in diameter	0.6–7
12-in diameter	0.4–4

Parking Area

How many acres are needed for a 220-car parking lot? Local streets will be used for access to half the lot, and access roads will be built for the other half.

$$A = 300C_S + 400C_L$$

A = area of parking, ? ft^2

C_S = number of cars served by street access, 110
C_L = number of cars served by access roads, 110

$$A = 300 \times 110 + 400 \times 110 = 77,000 \text{ ft}^2$$
$$1 \text{ acre} = 43,560 \text{ ft}^2$$
$$A = \frac{77,000}{43,560}$$
$$= 1.77 \text{ acres}$$

Refer to Table 2-2.

TABLE 2-2 DESIRABLE PARKING RATIOS FOR BUILDINGS

Building type served, ft²	Parking area per gross floor area, ft²
Shopping centers	2.0–2.3
Theatres, churches, assembly halls	1.4–1.9
Hotels and motels served primarily by automobiles	1.0–1.4
Hotels and motels served primarily by transportation other than automobiles	0.2–0.8
Offices served primarily by automobiles	0.9–1.1
Offices served primarily by mass transit	0.3–0.6
Apartments, condominiums	0.4–0.5

Parking Entrance Ramp Length

What is the minimum radius of a three-quarter circular driving ramp for a parking garage whose floor-to-ceiling distance is 10 ft 6 in? The ramp will have a roof over it. (Refer to Fig. 2-4.)

STEP 1. Find length of ramp.

$$L = \frac{100V - 10S}{S} + 20$$

L = length of ramp, ? ft
V = vertical distance between top and bottom of ramp, 10.5 ft

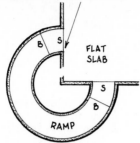

LENGTH & RADIUS OF RAMP ARE
MEASURED AT INSIDE OF CURB

FLAT
SLAB

RAMP

S = BEGINNING OF SLOPE
B = BREAKOVER ANGLE

FIGURE 2-4 Parking entrance ramp.

S = slope of ramp (from Table 2-1, maximum grade of roofed and curved parking access ramp), 12%

$$L = \frac{100 \times 10.5 - 10 \times 12}{12} + 20$$
$$= 97.5 \text{ ft}$$

STEP 2. Find radius of ramp at inside curb.

$$L = 2\pi RC$$

L = length of ramp, 97.5 ft
R = radius of ramp, **?** ft
C = circle fraction of ramp, three-quarters = 0.75

$$97.5 = 2 \times 3.14 \times R \times 0.75$$
$$R = 20.7 \text{ ft}$$

Sidewalk Width

A city sidewalk has a projected flow of 1200 people per hour during the busiest time of day. Along the street side of the sidewalk are parking meters, telephone poles, and other street furniture, and on the inside are store windows. How wide should the sidewalk be?

$$S = F + W + \frac{P}{400} + 3$$

S = sidewalk width, ? ft
F = width of street furniture (if located on pavement), 2 ft
W = window-shopping corridor (if store windows exist), 1.5 ft
P = pedestrian flow rate (maximum), 1200 people per hour

$$S = 2 + 1.5 + \frac{1200}{400} + 3$$
$$= 9.5 \text{ ft}$$

ARCHITECTURAL SPACE

This section includes methods of calculating a building's volume, area, and efficiency.

Building Volume

A building's volume includes the cubic footage (length × width × height) of every architectural space and the construction between the spaces, in the outer walls, under the lowest floor, and above the top floor. Volumes should be calculated as follows (see Fig. 2-5):

> *At full volume*   Rooms, walk-in penthouses, finished basements, dormers, bay windows, walk-through tunnels, tanks, vaults, balconies, enclosed porches, screened areas, chimney masses, and finished attics.

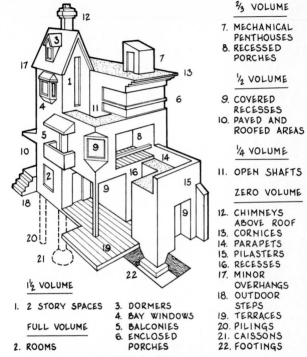

FIGURE 2-5 Building volumes.

At 0.67 volume Porches recessed into the building and not enclosed by sashes or screens, mechanical penthouses with lower than 6-ft 8-in headroom, and unfinished attic areas having at least 5-ft 0-in headroom and permanent staircase access, garages, and unfinished areas above ground level having doorway access.

At 0.5 volume   Covered recesses, nonenclosed roofed porches extending from the building, paved areas having building walls on two sides and a roof over, areaways, pipe tunnels, utility pits and trenches lined with construction materials, and unfinished attic areas having at least 5-ft 0-in headroom and movable staircase or hatch access, and unfinished basements.

At 0.25 volume   Open shafts, courtyards, and all other attic voids not counted above.

At zero volume   Recesses, courts, terraces, outside steps, parapets, garden walls, cornices, belvederes, chimneys above roofs, pilasters and other protrusions on exterior walls, cornices, crawl spaces, footings, pilings, and caissons.

Building Floor Area

A building contains two kinds of floor space: (1) net area and (2) gross area.

Net area is the sum of all usable, assignable, or programmable spaces and storage areas measured to the inside faces of the exterior walls. This includes walls between two assignable spaces but excludes walls between assignable spaces and general areas. Net area does not include mechanical, custodial, toilet, shaft, duct, elevator, stair, lobby, and general circulation areas.

Gross area is the total floor area of a building as measured from the outer face of the exterior walls. This includes bay windows, dormers, chimneys, tanks, vaults, shafts, and areas with a sloping ceiling having headroom of at least 5 ft under the lower end.

Floor areas should be calculated as follows (see Fig. 2-6):

At 1.5 area   Two-story areas. Heights in excess of two stories count as half their area per story.

At full area   All usable and livable spaces with an average headroom of at least 6 ft 8 in, including horizontal and vertical circulation, storage, utility, and mechanical areas.

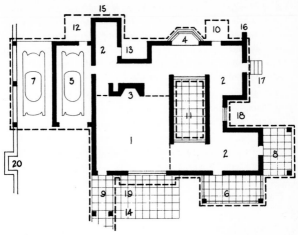

1½ AREA

1. 2 STORY SPACES

FULL AREA

2. FINISHED ROOMS
3. CHIMNEY
   MASSES
4. BAY WINDOWS

⅔ AREA

5. GARAGES & UN-
   FINISHED ROOMS

6. SCREENED
   PORCHES

½ AREA

7. CARPORTS
8. OPEN PORCHES
9. PAVED &
   ROOFED WALKS

¼ AREA

10. CANOPIES
11. COURTYARDS
12. LARGE
    OVERHANGS

13. ROOFED
    RECESSES
14. TERRACES

ZERO AREA

15. MINOR
    OVERHANGS
16. PILASTERS
17. OUTSIDE STAIRS
18. UNROOFED
    RECESSES
19. LIGHTWEIGHT
    SUNSCREENS
20. **GARDEN WALLS**

FIGURE 2-6 Building floor areas.

At 0.67 area   Garages, enclosed and screened porches, mechanical penthouses with lower than 6-ft 8-in headroom, unfinished walk-through tunnels, roofed balconies, and unfinished areas having doorway access.

At 0.5 area   Carports, paved and roofed walks, open porches, and entrance porticoes.

At 0.25 area   Canopies, paved or landscaped areas under large overhangs, courtyards, terraces, paved and unroofed areas that have building walls on at least three sides.

At zero area   Crawl spaces, pits and trenches below the bottom floor or outside the exterior wall, footings, foundations, pilasters and other protrusions on the exterior walls, recesses, attic areas having less than 5-ft headroom, minor roof overhangs, protruding sun-screens of light construction, chimneys above the roof, and all other unlivable areas.

Building Efficiency

A suburban library has a net area of 24,100 ft² and a gross area of 28,500 ft². What is the building's efficiency?

$$E = 100 \frac{N}{G}$$

E = building efficiency, ?%
N = net area of building, 24,100 ft²
G = gross area of building, 28,500 ft²

$$E = 100 \times \frac{24,100}{28,500} = 84.6\%$$

OCCUPANCY LOAD

Most commercial and residential architecture has a typical amount of floor space devoted to each user of the building. These areas vary widely according to the activity of each inhabitant and the function of the building.

Table 2-3 presents typical occupancy loads for today's architecture. These amounts may vary considerably as a result of specific architectural requirements or interspatial relationships.

TABLE 2-3 TYPICAL OCCUPANCY LOADS*

Use	Net floor area per occupant, ft^2	Grossing factor, × net area†
Apartment	250	1.25
Assembly areas (conference chambers, meeting rooms, small lecture areas)	15	2.0
Assembly areas, without individual seats (auction rooms, churches, chapels, lodge rooms, reviewing stands, stadiums)	7	2.5
Assembly areas, with individual seats (auditoriums, theatres, lecture halls):		
Movable seats	10	2.4
Fixed seating	11	2.2
Theatre stage apron (minimum 28-ft width and 6-ft depth): total area	250	1.8
Theatre stage (minimum 25 ft working depth, 28 ft proscenium width, 50 ft total width): total area	1500	1.5
Standing room for audience, places of assembly	4	1.5
Lobby and other public areas	10	1.2
Ticket booths: each	30	1.8
Athletic facilities:		
Billiard rooms	60	1.5
Bowling alleys	75	1.3

TABLE 2-3   TYPICAL OCCUPANCY LOADS* *(Continued)*

Use	Net floor area per occupant, ft^2	Grossing factor, $\times$ net area†
Dance floors	12	1.5
Gymnasiums	50	1.4
Skating rinks	65	1.4
Water or ice area	50	
Surrounding deck	15	
Locker rooms	60	1.6
Locker areas and inner circulation	35	
Showers and toilets	25	
Areas for training, first aid, towel service, storage	10	
Drinking establishments, seating and inner circulation:		
Taverns and bars	18	1.3
Night clubs	25	1.3
Food service, cafeterias: total	28	1.5
Eating (seating and inner circulation)	14	
Serving, vending, and disposal	6	
Kitchen, food storage, and administration	9	
Food service, luncheonettes: total	23	1.25
Eating	12	
Serving, vending, and disposal	4	
Kitchen, food storage, and administration	7	
Food service, restaurants: total	35	1.5
Eating	21	
Serving, vending, and disposal	5	
Kitchen, food storage, and administration	9	

TABLE 2-3 TYPICAL OCCUPANCY LOADS* *(Continued)*

Use	Net floor area per occupant, ft²	Grossing factor, × net area†
Hotels and motels: total	130	1.4
Private areas	120	
Lobby	10	
Hospitals, per bed unit:		
Infirmary areas	125	1.65
Psychiatric	165	1.6
Diagnostic and treatment	50	1.5
Services (food, maintenance, staff facilities)	45	1.3
Administration	20	1.5
Institutions (childrens' homes, sanitariums, nursing homes), sleeping areas:		
Infants	40	1.6
Children	70	1.6
Adults	100	1.6
Administration, treatment, and services	80	1.5
Industrial:		
Shops: work station and storage	200	1.4
Laboratories: station and storage	140	1.5
Libraries:		
Public:		
Reading rooms, per user	35	1.5
Stack space, per bound volume	0.08	1.3
University:		
Undergraduate reading rooms, per user	35	1.5
Graduate reading rooms, per user	60	1.4

TABLE 2-3   TYPICAL OCCUPANCY LOADS* *(Continued)*

Use	Net floor area per occupant, ft²	Grossing factor, × net area†
Stack space, per bound volume	0.1	
Service space	25% of reader space	1.5
Museums: circulation area around exhibits	15	1.2 × total exhibit area
Offices:		
Management areas	150	1.25
Nonmanagement areas	120	1.25
Service (lounge, coffee)	3	1.25
Retail:		
Basement	20	1.35
Ground floor	30	1.35
Upper floors	50	1.35
Schools:		
Small classrooms	20	1.65
Large classrooms	15	1.65
Instructional laboratories	70	1.65
Lecture halls	12	1.65
Seminar rooms	20	1.65
Shops and vocational rooms	50	1.65
Reading rooms	40	1.65
Dormitories: sleeping and lounge areas	100	1.8
Kindergarten: seating and play areas	50	1.8
Day-care nurseries: seating and play areas	40	2.0
Rest rooms: men or women	100 ft² + 1.2 ft² per occupant	

TABLE 2-3 TYPICAL OCCUPANCY LOADS* (Continued)

Use	Net floor area per occupant, ft²	Grossing factor, × net area†
Transportation:		
Auto parking: indoor	180	1.15
Passenger terminal loading platforms	1.5 × full vehicle unloading capacity	
Loading docks (outdoor parking):		
Loading and breakdown area, per bay	400	1.4
Security office, per employee	80	1.4
Security entrance: door, steps	100	
Trash holding room, per building occupant	0.3	1.5

*These amounts are not to be confused with the occupancy loads in the Uniform Building Code, which are minimum amounts.

†Grossing factor is the reciprocal of building efficiency × 100.

Example 1 A restaurant manager wants to open a cafe on the bottom floor of a building located on the corner of a busy downtown intersection. From the pedestrian traffic at the corner, the manager estimates he can draw 170 customers for lunch during two sittings. The total floor area available is 2500 ft². He desires to find the most profitable area for his cafe and then to sublease the rest of the space. What is the optimum floor area for the cafe and the remaining space he can sublet?

$$L = \frac{F}{AG}$$

L = occupancy load, 170 people ÷ 2 sittings = 85 people
F = optimal floor area, ? ft²

A = area per occupant (from Table 2-3, for luncheonette), 23 total net ft^2

G = grossing factor (from Table 2-3, for luncheonette), 1.25

$$85 = \frac{F}{23 \times 1.25}$$
$$F = 85 \times 23 \times 1.25 = 2444 \text{ ft}^2$$

Subleasable floor area equals $2500 - 2444 = 56$ ft^2. This is probably not enough area to plan subletting; thus the manager should consider using the whole bottom floor for the restaurant.

Example 2   What is the approximate area of a two-bay service loading facility for a commercial office building that accommodates 800 employees?

$$A = A_1 + A_2 + A_3 + \cdots A_n$$

A = total floor area of space, ? ft^2

$A_1, A_2, A_3, \ldots, A_n$ = floor areas of each partial space

An office loading facility of this kind consists of a loading and break-down area, a security entrance with a small office for one guard, and a trash holding room. Here, the grossing factor is ignored because the problem involves net area only. From Table 2-3:

Loading and breakdown area, 400×2 bays = 800 ft^2
Security entrance, $100 + 80 = 180$ ft^2
Trash holding area, 0.3×800 employees = 240 ft^2

$$A = 800 + 180 + 240$$
$$= 1220 \text{ net square feet}$$

OCCUPANCY FLOW

Certain architectural spaces used primarily for circulation must have dimensions that ensure the comfort and safety of its pedestrian users.

This section takes into consideration requirements of the American National Standards Institute (ANSI).

Stairs

Below are the Uniform Building Code (UBC) requirements for stair steps.

Use	Riser height, in	Tread width, in
Public	4 to 7½	10 or more
Private	4 to 8	9 or more
Ideal	7	11

Number of steps   A commercial staircase with a halfway landing is to be constructed between two floors whose finished surfaces are 11 ft 3 in apart. How many risers should the staircase have?

$$R \approx \frac{D}{7}$$

R = number of risers in staircase, ? units
D = vertical distance between top and bottom of staircase, 11 ft 3 in
   = 135 in

$$R \approx \frac{135}{7} = 19.3$$

Use either 19 or 20. Because the landing occurs at midpoint, use an even number.

$$R = 20 \text{ risers}$$

Riser height   A granite staircase in a city park is to have treads 18 in wide. What is the ideal riser height?

$$R = 9 - \sqrt{\frac{(T - 8)(T - 2)}{7}}$$

R = Riser height, ? in
T = Tread width, 18 in

$$R = 9 - \sqrt{\frac{(18 - 8)(18 - 2)}{7}} = 4.22 \text{ in}$$

Tread width   The risers in a residential staircase must be $7^{11}\!/_{16}$ high. What is the ideal tread width?

$$T = 5 + \sqrt{7(9 - R)^2 + 9}$$

T = tread width, ? in
R = riser height, $7^{11}\!/_{16}$ in = 7.69 in

$$T = 5 + \sqrt{7(9 - 7.69)^2 + 9} = 9.59 \text{ in}$$

Elevators
Below are UBC and ANSI requirements for elevators. (Refer to Fig. 2-7.)

Dimension	Requirement
Minimum area per person	2 ft²
Maximum live load capacity	120 lb/ft²
Minimum depth	54 in
Minimum width:	
Capacity less than 2000 lb	54 in
Capacity 2000 lb or more	68 in
Minimum area:	
Capacity less than 2000 lb	20.25 ft²
Capacity 2000 lb or more	25.5 ft²
Minimum opening width	36 in
Desirable opening width	42 in
Desirable shaft area	Approx. 1.7 of cab area

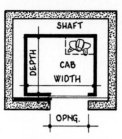

FIGURE 2-7   Elevator.

Elevator speed   How fast should an elevator travel in a building that is 240 ft tall?

$$S = 1.6H + 350$$

S = speed of elevator, ? ft/min
H = height of building, 240 ft

$$S = 1.6 \times 240 + 350 = 734 \text{ ft/min}$$

NOTE: Because of the numerous stops an elevator usually makes, it typically has a net velocity of about 0.25 maximum velocity during peak traffic periods.

Elevator size   An eight-story office building whose space is leased by several different businesses has 6870 ft² of rentable floor space on each level. If the building has two elevators, what should the cab area of each elevator be?

$$226NE = AFC$$

N = number of elevators, 2 units
E = area of elevator cab, ? ft²

A = net area of each floor, 6870 ft^2
F = number of floors in building, 8
C = elevator capacity factor (from Table 2-4, for a diverse-occupancy commercial building), 0.22

$$226 \times 2 \times E = 6870 \times 8 \times 0.22$$
$$E = \frac{6870 \times 8 \times 0.22}{226 \times 2} = 26.8 \text{ ft}^2$$

TABLE 2-4   ELEVATOR CAPACITY FACTORS

Building type	Capacity factor
Office:	
Single-purpose occupancy	0.33
Diverse occupancy	0.22
Apartment	0.15
Hotel	0.38
Hospital	0.12
Parking garage	0.16

Escalator Capacity

An escalator in a department store is 32 in wide and moves at 120 ft/min. How many people per hour can the escalator carry?

There are only four possible escalator capacities. They are:

1. 32 in wide at 90 ft/min: 5000 persons per hour
2. 32 in wide at 120 ft/min: 6250 persons per hour
3. 48 in wide at 90 ft/min: 8000 persons per hour
4. 48 in wide at 120 ft/min: 10,000 persons per hour

NOTE: The standard slope of an escalator is 30°. The treads of each step are 13 in wide, and the risers are 8 in high.

Ramp Length

A ramp is to be built from the concrete floor of a building entrance to the surrounding ground 6 ft 2 in below. How long should the ramp be? The amount should include the required length for landings.

$$L = \frac{H}{0.083} + \left[\frac{H}{2.49} \right] \times 5$$

L = length of ramp, including landings, ? ft
H = height of ramp, 6 ft 2 in = 6.17 ft

$$L = \frac{6.17}{0.083} + \left[\frac{6.17}{2.49} \right] \times 5$$
$$= 74.3 + [2.47] \times 5 = 84.3 \text{ ft}$$

Distance between Two Exits

The floor plan of an L-shaped building 80 ft wide and 112 ft long requires two fire stairs. How far apart should they be?

$$D \geq \frac{M}{2}$$

D = distance between centers of exit doors, ? ft
M = maximum diagonal length of floor plan:

$$\sqrt{L^2 + W^2} = \sqrt{80^2 + 112^2} = 138 \text{ ft}$$
$$D \geq \frac{138}{2} = 69 \text{ ft minimum}$$

NOTE: This minimum distance is measured from the centers of the doorways of the two staircases.

SIGNAGE

Semantic communication is often required to guide persons through architectural sites and spaces. The relationship between the height of the words and the distance of the approaching viewer must allow a sign to be readable to someone having 20/40 vision. In cases involving vehicular traffic, the speed of the approaching viewer must also be considered.

Messages that begin with an uppercase letter followed by lowercase characters are usually more readable than those that contain uppercase letters only. Letter height is that of uppercase letters.

Pedestrian Signage

The name of a building is to be mounted on the rear wall of the entrance lobby. If the letters are 46 ft away from the entrance and should be readable from this point, how high should they be?

$$H = \frac{D}{25}$$

H = height of readable letters, ? in
D = distance between pedestrian viewer and sign, 46 ft

$$H = \frac{46}{25} = 1.84 \text{ in}$$

Vehicular Signage

The owner of a roadside restaurant wants to erect a large sign in plain view of travelers, who drive by at about 55 mi/hr. How high should the letters of the most important words in the sign be so that an oncoming driver can have time to read the sign, slow down, and enter the restaurant?

$$H = 0.04(A + T)$$

H = height of readable lettering, **?** in
A = apprehension and reaction time, 3 × speed (S) of approaching vehicle (55 mi/hr), in seconds; $A = 3S$ sec
T = slowdown and stopping time = $S^2/9$ (For vehicular traffic signs that do not require stopping, such as directional and mileage signs, T may be ignored.), in seconds; $T = S^2/9$ sec
S = speed of traveler, 55 mi/hr

$$H = 0.04 \left(3S + \frac{S^2}{9} \right)$$

$$= 0.04 \left(3 \times 55 + \frac{55^2}{9} \right) = 20.0 \text{ in high}$$

OPENINGS

Certain openings in a building envelope must be of minimum size in order to ensure the durability of the architecture and the comfort of its occupants.

Fenestration Area

How much window area should a living room 21 by 15 ft in size have?

$$A \geq \frac{F}{10}$$

A = minimum area of unobstructed glass, not including frame, **?** ft^2
F = floor area of room, 21 × 15 = 315 ft^2

$$A \geq \frac{315}{10} = 31.5 \text{ ft}^2 \text{ minimum}$$

Openable Fenestration Area

A homeowner wishes to build a leanto greenhouse against the south side of her house. The space will be 14 ft wide and 38 ft 8 in long. How much openable glass should this room have?

$$G \geq \frac{F}{20}$$

G = minimum area of openable glass, ? ft^2
F = floor area of room having no mechanically operated ventilating system, $14 \times 38.7 = 542$ ft^2

$$G \geq \frac{542}{20} = 27.1 \text{ ft}^2 \text{ minimum}$$

Attic Vent Size

The enclosed rafter spaces under the gable roof of a residence measure 44 ft long by 28 ft wide. How large should the eave vents at each end of the roof be?

$$V \geq \frac{F}{150}$$

V = minimum open area of vents not including frame, ? ft^2
F = floor area of attic, $44 \times 28 = 1232$ ft^2

$$V \geq \frac{1232}{150} = 8.21 \text{ ft}^2 \text{ total area}$$

Under the gable roof will be a vent at each end, for two vents total.

$$\text{Area of each vent} = \frac{8.21}{2}$$
$$= 4.11 \text{ ft}^2 \text{ minimum}$$

Crawlspace Vent Size

The crawlspace of a residence 44 ft long and 28 ft wide requires how many standard foundation wall vents mounted in its top course?

$$A \geq \frac{P}{50}$$

A = net free open area of vent not including frame, ? ft^2
P = perimeter length of foundation wall, $2(44 + 28) = 144$ ft

$$A \geq \frac{144}{50} = 2.88 \text{ ft}^2 \text{ minimum free open area of vent}$$

Number of vents: A standard 8- by 16-in corrosion-resistant foundation vent contains 0.45 ft^2 of free open area.

$$\text{Number of vents} = \frac{2.88}{0.45} = 6.4$$

Use at least seven.

NOTE: Crawlspace vents should be mounted opposite each other and near corners as much as possible. Every enclosed crawlspace area should contain at least four standard vents or equal.

PLANNING GEOMETRY

Use of certain geometric relations may lead to ease of architectural planning and construction and ultimately to greater architectural beauty and order. The merits of such patterning relate strongly to the inclinations of the designer.

The Octagonal Square

By using the diagram shown in Fig. 2-8, designers and builders can create architectural forms that possess the construction simplicity of the square and the fluidity of the curve.

What are the outline dimensions of each part of the octagonal seating arrangement shown in Fig. 2-9? The outer and inner diameters are 134 in and 70 in, respectively.

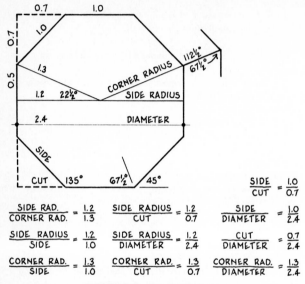

$$\frac{SIDE}{CUT} = \frac{1.0}{0.7}$$

$$\frac{SIDE\ RAD.}{CORNER\ RAD.} = \frac{1.2}{1.3} \qquad \frac{SIDE\ RADIUS}{CUT} = \frac{1.2}{0.7} \qquad \frac{SIDE}{DIAMETER} = \frac{1.0}{2.4}$$

$$\frac{SIDE\ RADIUS}{SIDE} = \frac{1.2}{1.0} \qquad \frac{SIDE\ RADIUS}{DIAMETER} = \frac{1.2}{2.4} \qquad \frac{CUT}{DIAMETER} = \frac{0.7}{2.4}$$

$$\frac{CORNER\ RAD.}{SIDE} = \frac{1.3}{1.0} \qquad \frac{CORNER\ RAD.}{CUT} = \frac{1.3}{0.7} \qquad \frac{CORNER\ RAD.}{DIAMETER} = \frac{1.3}{2.4}$$

FIGURE 2-8   Octagonal square.

STEP 1.   Find outer edge length from the octagonal square ratio:

$$\frac{\text{Side}}{\text{Diameter}} = \frac{1.0}{2.4}$$

Side = outer side; diameter = 134 in

$$\frac{\text{Outer side}}{134} = \frac{1.0}{2.4}$$

$$\text{Outer side} = \frac{134 \times 1.0}{2.4} = 55.8 \text{ in}$$

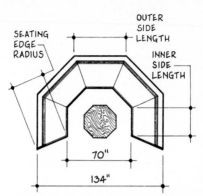

FIGURE 2-9 Octagonal seating arrangement.

STEP 2. Find the inner side length from the same ratio.

$$\frac{\text{Side}}{\text{Diameter}} = \frac{1.0}{2.4}$$

Side = inner side; diameter = 70 in

$$\frac{\text{Inner side}}{70} = \frac{1.0}{2.4}$$

$$\text{Inner side} = \frac{70 \times 1.0}{2.4} = 29.2 \text{ in}$$

STEP 3. Find the seating edge radius from the octagonal square ratio:

$$\frac{\text{Side radius}}{\text{Corner radius}} = \frac{1.2}{1.3}$$

Side radius = seat depth = $\dfrac{134 - 70}{2}$ = 32 in

Corner radius = seating edge radius

$$\frac{32}{\text{Seating edge radius}} = \frac{1.2}{1.3}$$

$$\text{Seating edge radius} = \frac{32 \times 1.3}{1.2} = 34.7 \text{ in}$$

The Golden Section

An architect desires to give a rectangular doorway ideal proportions relating to the Golden Section. If the doorway's height is 78 in, what should its width be?

$$\frac{A}{B} = 0.618 = \tan 31.72°$$

A = horizontal dimension of doorway, **?** in
B = vertical dimension of doorway, 78 in

$$\frac{A}{78} = 0.618$$
$$A = 0.618 \times 78$$
$$= 48.2 \text{ in}$$

The Modified Modular

Le Modular, with its two columns of "Red" and "Blue" series, was invented by Le Corbusier in 1948. Despite the harmonious compositions and relation to human movement that evolves from its use, the Modular has found scant use in this country because of its seeming incompatibility with U.S. manufacturing and construction procedures. However, in 1970 this author added to Le Corbusier's Red and Blue series columns of "Pink" and "Purple" series, then rounded all large numbers to multiples of 2. He has found this new mathematic matrix to relate to American technology (see Table 2-5). Of the new numbers, 24 and 48 in are important modular dimensions of American building materials, 30 in is the standard American table height, 38 in is a comfortable single bed and private hallway width, 62 in is a good public corridor width, and 78 in is a good door and window header height. In these ways, the Modified Modular brings together aesthetics, comfort, and economics.

When utilizing this system, one does not need to employ exact numbers. Tolerances of up to 2 percent allow for flexible use without altering the perception of proportions the system creates.

TABLE 2-5 THE MODIFIED MODULAR, in inches

Pink	Red	Blue		Purple
		1		
		2		
		3		
		5		
	4	8		18
14	(6)	6½	⑬ →	30
24	(10)	10½	㉑ (20)	48
38 ←	(16)–	17	34	78
62	(28)	27½	55 (54)	
100	(44)	44½	89 (88)	
		72	144	

Dimensioning with the modified modular What are comfortable dimensions for a dining area for six people?

STEP 1. Consult Table 2-6 for appropriate dimensions for each activity occurring in the dining area.

STEP 2. Lay out in sketch plan or section the dimensions for each activity.

STEP 3. Add the numbers each way to find the area's length and width.

TABLE 2-6 MODULAR HUMAN ACTIVITY DIMENSIONS

Activity	Dimension, in
Individual use area for one: at table, desk, vanity, or counter	14 × 24
Individual use access border, along back or sides of horizontal use area	10–14 wide
Counter depth: use area plus access border	24 or 28 deep
Front-to-rear distance required for standing person or person seated at table or counter	17
Single walking person's width	24 at shoulders, 17 below thighs
Aisle, residential	38
Aisle, commercial (ANSI minimum is 60 in)	62
Bed size:	
Single person	28–38 wide, 72 or 78 long
Two people	54 or 62 wide, 72 or 78 long
Toilet use area	30 × 54

TABLE 2-6   MODULAR HUMAN ACTIVITY DIMENSIONS *(Continued)*

Activity	Dimension, in
Leg room and access width, such as between seating area and coffee table	21
Shelf distance, from front of body to center of shelf	34
Chair height	17
Table height	28 or 30
Counter height	34 or 38
Door and window head height	78
Window sill height:	
Standing adult behind	48 or 55
Seated adult behind	34 or 38
Eye level of seated adult, above floor	44–48
Seating size:	
Dining chair	17 × 17
Sofa cushion unit	24 × 24

Two possible dining arrangements are shown in Figs. 2-10 and 2-11.

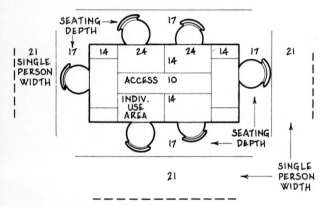

LENGTH OF DINING AREA
    21 + 17 + 14 + 24 + 24 + 14 + 17 + 21 = 152" = 12'-8"

WIDTH OF DINING AREA
    38 + 17 + 14 + 10 + 14 + 17 + 21 = 131" = 10'-11"

FIGURE 2-10 Dining room arrangement 1.

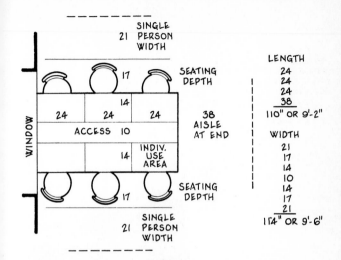

FIGURE 2-11 Dining room arrangement 2.

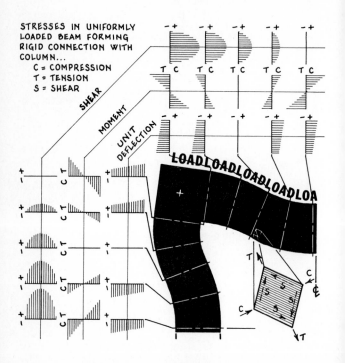

STRESSES IN UNIFORMLY
LOADED BEAM FORMING
RIGID CONNECTION WITH
COLUMN...
    C = COMPRESSION
    T = TENSION
    S = SHEAR

SHEAR

MOMENT

UNIT DEFLECTION

LOADLOADLOADLOADLOADLOA

Mechanics of uniform load on beam and post.

STRUCTURE

GENERAL

Architecture is an assembly of structural parts. Each piece must be able to support itself, sometimes the weight of other structural members, often a portion of the occupants and furnishings in the architecture, and almost always a share of the climatic forces that act upon the total mass of materials.

This interplay of architectural forces upon each structural member is described in an equation. The mathematics on one side of the equal sign defines the *load* acting upon the structure; the mathematics on the other side describes the *structure* that resists the load.

Load = structure

| *Load:* Weight and arrangement of people, furnishings, building materials, and climatic forces | = | *Structure:* Size, shape, material, and position of the structural part |

$$\frac{WL}{8} = S\,\frac{bd^2}{6}$$

The following section deals with the load side of structural problems; subsequent sections concern the structural side in terms of various architectural materials.

When performing structural calculations, always consult applicable local codes that may control for specific design.

LOADS

Load analysis involves the following factors:

1. Dead loads
2. Live loads

3. Moving loads
4. Load distributions

Some *general load requirements* are given below.

1. Garage floors used for parking private pleasure autos shall each be designed to support concentrated wheel loads of not less than 2000 lb/ft^2 spread over a 2½-ft^2 area.

2. Greenhouse roof bars, purlins, and rafters shall each be designed to support a 100-lb concentrated load per 12 ft of length in addition to all other loads.

3. All roofs shall have sufficient slope or camber to assure adequate drainage after long-term deflection from dead load has been attained.

4. On sloping surfaces, live loads shall be assumed to act vertically on a horizontal projection of the sloping surface area.

5. The weight of drained earth behind a retaining wall is considered as equal to a fluid weighing 30 lb/ft^2 and having a depth equal to the depth of the retained earth. Any surcharge present behind the walls must be calculated as an additional load.

6. Walls exceeding 6 ft in height shall be designed to withstand a lateral load of not less than 5 lb/ft^2. The deflection of such walls shall not exceed 1/240 of the span if the finish is plaster or some other brittle material or 1/120 of the span if the finish is wood or some other resilient materials.

Dead Loads

Dead load (DL) is the weight of the architecture resting upon the structural member. It is considered to be ever-present and unmoving. Typical weights of architectural construction types are listed in Table 3-1.

TABLE 3-1 TYPICAL WEIGHTS OF CONSTRUCTION TYPES

Construction	Weight, lb/ft^2 surface area
Wood stud wall, 2 x 4, interior, ½-in drywall 2S	8
Interior, wood or metal 2 x 4s, plaster 2S	19
Exterior, drywall; 4-in batt insul.; wood siding	11
Exterior, drywall; 4-in batt insul.; 4-in brick (MW)	47
Exterior, drywall; 4-in batt insul.; 8-in concrete block	60–65
Metal stud wall, 2 x 4, interior, ½-in drywall 2S	7
Exterior, drywall; 4-in batt insul.; 1-in stucco	23
Metal stud wall, exterior, drywall; 4-in batt insul.; 2-in dryvit	18
Exterior, drywall; 4-in batt insul.; 3-in granite or 4-in brick	55
Plaster, per face, wall, or ceiling, on masonry or framing	8
Ceramic tile veneer, per face	10
Masonry wall, 4-in brick, MW, per wythe	39
4-in conc. block, heavy aggregate, per wythe	30
8-in conc. block, heavy aggregate, per wythe	55
Glass block wall, 4-in thick	18
Glass curtain wall	10–15
Floor or ceiling, 2 x 10 wood deck, outdoors	8–10
Wood frame, 2 x 10, interior, unfinished floor; drywall ceiling	8–10
Concrete flat slab, unfinished floor; susp. ceiling	80–90
Concrete pan joist (25 in o.c., 12-in pan depth, 3-in slab), unfinished floor; susp. ceiling	90–100
Concrete on metal deck on steel frame, unfinished floor; susp. ceiling	65–70

TABLE 3-1 TYPICAL WEIGHTS OF CONSTRUCTION TYPES *(Continued)*

Construction	Weight, lb/ft² surface area
Finished floors, add to above:	
Hardwood	3
Floor tile	10
1½-in terrazzo	25
Wall-to-wall carpet	2
Roof, sloping rafters or timbers, sheathing; 10-in batt insul.; ½-in drywall	12–15
Built-up 5-ply roofing, add to above	6
Metal roofing, add to above	3–4
Asphalt shingle roofing, add to above	4
Slate or tile roofing, ¼ in thick, add to above	12
Wood shingle roofing, add to above	3–5
Insulation, batt, per 4-in thickness	1
Insulation, rigid foam boards or fill, per inch thickness	0.17
Stairways:	
Concrete	80–95
Steel	40–50

Example 1 The main floor of a town public library is carpet on concrete slab on steel deck on W 16 x 36s, with a suspended ceiling of acoustic tile underneath. If the typical bay size is 12 ft 4 in wide and 26 ft 0 in long, what is the dead load resting on each girder?

$$DL = AW$$

DL = total dead load supported by each girder, ? lb

A = area in square feet of floor supported by the structural member (each girder supports half the floor area between it and the next girder on each side), $12.3/2 \times 26 \times 2 = 320$ ft² (See Fig. 3-1.)

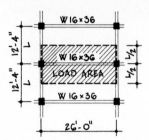

FIGURE 3-1 Dead load on girder 1.

W = weight of the floor in pounds per square foot (as listed in Table 3-1, W for floor, concrete on metal deck on steel frame, finished floor/suspended ceiling + wall-to-wall carpet = (65 to 70) + 2), 70 lb/ft² wall area

$$DL = 320 \times 70$$
$$= 22,400 \text{ lb}$$

Example 2 In the town library from Example 1 above, the W 16 x 45 end girders on the long side of the 12 ft 4 in by 26 ft 0 in structural bays each supports in addition to part of the floor an exterior wall of masonry 14 ft 8 in tall. If the wall is plaster on 8-in concrete block on 2-in rigid insulation with an exterior finish of 4-in brick, what is the total dead load resting on an end girder?

$$DL = A_1 W_1 + A_2 W_2 + \cdots A_n W_n$$

DL = total dead load supported by the structural member, ? lb

A_1 = area in square feet of floor supported by the structural member (the end girder supports half the floor area between it and the next girder), $12.3/2 \times 26$ = 160 ft² (See Fig. 3-2.)

W_1 = weight of the floor supported by the girder in pounds per square foot (as listed in Table 3-1), 70 lb/ft²

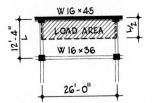

FIGURE 3-2 Dead load on girder 2.

A_2 = surface area in square feet of the wall supported by the girder equals height of wall × length of girder, $14.3 \times 26 = 372 \text{ ft}^2$

W_2 = weight of the wall supported by the girder in pounds per square foot (as listed in Table 3-1), $8 + 55 + 2(0.17) + 39 = 102 \text{ lb}/\text{ft}^2$

$$DL = 160 \times 70 + 373 \times 102$$
$$= 49,100 \text{ lb}$$

Live Loads

Allowable live loads (LL) for common occupancies are listed in Table 3-2.

TABLE 3-2 ALLOWABLE LIVE LOADS

Occupancy	Uniform load, lb/ft²	Concentrated load, lb/ft²
Apartments, residences, hotels, clubs, hospitals, prisons	40	
Armories	150	
Auditoriums, theatres, dance halls, gyms, plazas, terraces, play areas:		
Fixed seating	50	
Movable seating and circulation	100	
Stage areas and platforms	125	

TABLE 3-2 ALLOWABLE LIVE LOADS *(Continued)*

Occupancy	Uniform load, lb/ft²	Concentrated load, lb/ft²
Exit corridors, stairways, balconies, fire escapes	100	
Garages, general storage or repair	100	2000
Private pleasure car storage	50	2000
Libraries:		
Reading rooms	60	1000
Stack rooms	125	1500
Manufacturing:		
Light	75	2000
Heavy	125	3000
Offices	50	2500
School classrooms	50	1000
Sidewalks and driveways	250	2000
Storage:		
Light	125	
Heavy	250	
Stores:		
Retail	75	2000
Wholesale	100	3000
Roofs, general, unless otherwise noted	30	
Floors supporting movable walls	add 20 lb/ft²	

Live load reduction: floors How much live load reduction is allowed for a W 16 x 36 girder that supports 320 ft² of public library floorspace if the dead load equals 75 lb/ft² and the live load equals 60 lb/ft²?

STRATEGY Solve for the live load reduction in the three formulas below and let the smallest percentage govern.

1. $R = 0.08A$

2. $R = 40\%$

3. $R = 23.1 \left(1 + \dfrac{DL}{LL}\right)$

R = live load reduction, ?%

A = area of floor in square feet (this reduction may be taken only if $A = 150 \text{ ft}^2$ or more and A is not used for assembly purposes), 320 ft^2

DL = dead load in pounds per square foot, 75 lb/ft^2

LL = live load in pounds per square foot (this reduction may be taken only if $LL = 100 \text{ lb/ft}^2$ or less), 60 lb/ft^2

1. $R = 0.08 \times 320 = 25.6\%$

2. $R = 40\%$

3. $R = 23.1 \left(1 + \dfrac{75}{60}\right) = 52.0\%$

Live load reduction = smallest percentage = 25.6%.

Live load reduction: roofs What is the allowable live load for a girder supporting 280 ft^2 of roof if its dead load is 35 lb/ft^2 and its pitch is 5 in 12?

Consult Table 3-3. The answer is 14 lb/ft^2.

TABLE 3-3 MINIMUM ROOF LIVE LOADS*

Roof slope	Roof area, ft²		
	0–200	201–600	600+
Flat or less than 4 in per foot	20	16	12
4 in to less than 12 in per foot	16	14	12
12 in or more per foot	12	12	12
Greenhouse or lath roofs	10	10	10
Awnings, if not cloth	5	5	5

*Where snow loads occur, the live load may be determined by the local building official.

Snow load reduction: roofs   If the allowable snow load for a flat roof is 30 lb/ft^2, what is the allowable snow load for a roof whose pitch is 9 in 12?

$$A = S - (P - 20)\left(\frac{S - 20}{40}\right)$$

A = allowable snow load, **?** lb/ft^2

S = allowable snow load for flat roofs according to local code (this reduction may be taken only if S = 20 lb/ft^2 or more), 30 lb/ft^2

P = pitch of roof in degrees (this reduction may be taken only if P = 20° or more), 9 in 12 = $\tan^{-1} 9/12$ = 36.9°

$$A = 30 - (36.9 - 20)\left(\frac{30 - 20}{40}\right)$$

$$= 25.8 \text{ lb/ft}^2 \text{ minimum}$$

Moving Loads

Moving loads include impact loads, crane runway loads, moving vehicular loads on bridges, special loads on elevator supports, loads resulting from moving parts in machinery, fatigue resulting from cyclic (repetitious) loading, and loads caused by vibration and oscillation. See Table 3-4 for moving load factors.

TABLE 3-4   MOVING LOAD FACTORS

Load	Load capacity increase factor
Elevator supports, connections, and cables	1.00
Crane runway supports and connections	0.25
Shaft- or motor-driven machinery, supports, and connections	0.20
Reciprocating machinery and connections	0.50
Moving loads in which maximum speed of load is known, when W = weight of load and V = velocity in miles per hour (1 mi/hr = 1.47 ft/sec)	$L = \dfrac{wv^2}{32}$

Example 1: Moving Concentrated Load   A ceiling girder in a factory has mounted on its flanges a 3-ton capacity moving hoist that weighs 85 lb. What is the hoist's design load on the girder? Where should the load be located?

$$P = W(1 + F)$$

P = total design load of the hoist on the girder, ? lb
W = weight of crane and its maximum load, $85 + 6000 = 6085$ lb
F = load increase factor (from Table 3-4, F for crane runway
        support girder), 0.25

$$P = 6085(1 + 0.25)$$
$$= 7607 \text{ lb}$$

Location of load: For purposes of calculation, a movable load is always located where it will create the maximum shear, moment, or deflection in its support. The same holds true for multiple loads.

Example 2: Impact Load   The railings of a reinforced concrete parking garage must be designed to contain a runaway car weighing 8000 lb and traveling at 12 mi/hr. If the center of load impact on the 40-in tall railing is 24 in above the floor, what is the total horizontal force at the railing's base? The perimeter of the floor is 24 in deep.

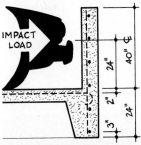

FIGURE 3-3   Impact load.

STRATEGY Take vertical moments at the level of bottom reinforcing in the rail, as shown in Fig. 3-3.

$$Pa = Rb$$

P = maximum force of car against rail; from Table 3-4, force of moving load, $wv^2/32$:

    w = weight of load, 8000 lb

    v = velocity of load, 12 mi/hr

    $P = 8000 \times 12^2/32 = 36{,}000$ lb

    If the car is 6 ft wide, P per linear foot is $36{,}000/6 = 6000$ lb/lin ft of width.

a = moment arm of load force from center of load to center of bottom reinforcing, $24 + 24 - 3 = 45$ in

R = resistance of structure at base, ? lb

b = moment arm of resisting force (because this force is in tension, the maximum force should not be computed at surface of concrete floor but at level of reinforcing 2 in below), b = distance from floor reinforcing to bottom reinforcing = $24 - 2 - 3 = 19$ in

$$6000 \times 45 = R \times 19$$
$$R = 14{,}200 \text{ lb/lin ft}$$

Example 3: Resonant Forces A 130-ft long, 12-ft wide reinforcing concrete mezzanine bridge overlooking the lobby of a hotel has a cross section as shown in Fig. 3-4. If the bridge is packed with people dancing to rock music, is the bridge safe? The unit stress of the concrete (f_c) is 3000 lb/in².

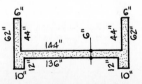

FIGURE 3-4 Cross section through concrete shape.

STEP 1.   Calculate the unit impact load on the bridge floor. People dancing to 4/4 music can be expected to create an impact load equal to twice their weight.

$$L = 2W_1$$

L_1 = live load caused by impact, ? lb/ft^2

W_1 = weight of people on the bridge floor (packed conditions indicate about 2 ft^2 per person at an average weight of 140 lb per person), $140/2 = 70$ lb/ft^2

$$L_1 = 2 \times 70 = 140 \text{ lb/ft}^2$$

STEP 2.   Compute the unit dead load of the bridge.

$$DL = \frac{AW_2}{144w}$$

DL = unit dead load of the bridge, ? lb/ft^2

A = cross-sectional area of the bridge section, $2 \times 62 \times 6 + 2 \times 4 \times 18 + 6 \times 136 = 1704$ in^2

W_2 = unit weight of concrete, 145 lb/ft^3

w = width of floor in feet, 12 ft

$$DL = \frac{1704 \times 145}{144 \times 12}$$
$$= 143 \text{ lb/ft}^2$$

STEP 3.   Determine the period of vibration for the mezzanine bridge.

$$T = 0.32 \sqrt{\Delta}$$

T = period of vibration for the construction, ? sec

Δ = unit deflection for the construction (the load condition is uniform load on beam with fixed ends):

$$\Delta = \frac{W_3 L_2^3}{384EI} \quad \text{(from Table 3-5)}$$

W_3 = weight of load; floor area (impact live load + dead load) = $12 \times 130\,(140 + 137) = 432{,}000$ lb

L_2 = length of span in inches, $130 \times 12 = 1560$ in

E = modulus of elasticity for concrete (from Table 3-20, E for concrete having $f_c = 3000$ lb/in^2), 3,150,000 lb/in^2

I = moment of inertia of the bridge cross section (this has been solved in the section on Concrete below, under "Moment of Inertia of Irregular Cross Section," page 178), 369,000 in^4

$$\Delta = \frac{432{,}000 \times 1560^3}{384 \times 3{,}150{,}000 \times 369{,}000} = 3.66 \text{ in}$$
$$T = 0.32\sqrt{3.62}$$
$$= 0.61 \text{ sec}$$

STEP 4.   Compare the bridge's *oscillation effect due to repeated rhythmic impact load* to the *damping effect due to dead load*. If this ratio is less than 1, the bridge is safe and no further analysis is necessary. If the ratio exceeds 1, a possibility of collapse exists and further investigation is required.

$$\frac{\text{Oscillation effect due to repeated impact load}}{\text{Damping effect due to dead load}} = \frac{L_1}{D} = ?$$
$$L_1 = 140 \text{ lb/ft}^2 \qquad D = 137 \text{ lb/ft}^2$$
$$\frac{140}{137} = 1.04$$

Possibility of collapse exists.

NOTE: Where the above ratio exceeds 1, special bracing or oscillation-damping are necessary.

STEP 5.   Compare the bridge's period of vibration to the beat of the music. Medium-fast rock music has a rhythm of about 100 beats per minute.

$$N = \frac{60}{T}$$

N = number of vibrations per minute of the bridge, **?**
T = period of vibration for the bridge (from step 3), 0.61 sec

$$N = \frac{60}{0.61} = 98.4 \text{ vibrations per minute}$$

If the music creates exactly the same number of beats per minute over an extended period of time, a danger of collapse exists. As the music has about 100 beats per minute, this is a definite possibility.

Summary   Under the conditions described above, the mezzanine bridge is in serious danger of collapsing. Because the repeated rhythmic impact load exceeds the dead load by 3 lb, every time the bridge oscillates in exact unison to the beats of the music an extra 3 lb/ft^2 of effective weight could be added to the floor. After only 60 sec a surplus load of 180 lb/ft^2 could exist on the already heavily crowded bridge. This would most likely make it collapse. Admittedly, this is a long shot based on several ballpark estimates, but it could happen.

Load Distribution
The way a load is distributed upon a structural member greatly affects the size of that structural member.

Vectors

Example 1   How many pounds tensile force exist in cable A? Does the force in cable B equal that in A? (See Fig. 3-5.)

STEP 1.   Find angle a.

$$\tan a = \frac{1}{10}$$
$$a = 5.71°$$

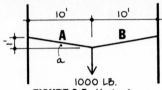

FIGURE 3-5 Vector 1.

STEP 2. Find A. Because of symmetry, half the load (or 500 lb) is held up by cable A and half by cable B.

$$\sin a = \frac{L}{A}$$

a = angle formed by cable A and the horizontal, 5.71°
L = weight of load held by cable A, 1000/2 = 500 lb
A = tensile force in cable A, ? lb

$$\sin 5.71° = \frac{500}{A}$$

$$A = \frac{500}{0.0955} = 5020 \text{ lb}$$

Does B equal A? Yes.

Example 2 How many pounds tensile force exist in cable A? (See Fig. 3-6).

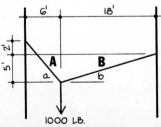

FIGURE 3-6 Vector 2.

Horizontal forces:

$$A \cos a = B \cos b$$

Vertical forces:

$$A \sin a + B \sin b = 1000$$

A = tensile force in cable A, ? lb
a = angle formed by cable A and the horizontal, tan a = $\frac{7}{6}$, a = 49.4°
B = tensile force in cable B, ? lb
b = angle formed by cable B and the horizontal, tan b = $\frac{5}{18}$, b = 15.5°

Horizontal forces:

$$A \cos 49.4° = B \cos 15.5°$$
$$A \times 0.651 = B \times 0.964$$
$$A = 1.48B$$

Vertical forces:

$$1.48B \times \sin 49.4° + B \sin 15.5° = 1000$$
$$1.48 \times 0.759B + 0.267B = 1000$$
$$B = 719 \text{ lb}$$
$$A = 1.48 \times B = 1.48 \times 719 = 1064 \text{ lb}$$

Example 3 A 200-lb man is standing 14 ft up on a 24-ft long aluminum ladder that weighs 35 lb (see Fig. 3-7). What is the horizontal force of the ladder against the pavement it is standing on? If the coefficient of static friction between aluminum and smooth concrete is 0.35, is the bottom of the ladder in danger of sliding?

STEP 1. Find the vertical distance between the pavement and point A.

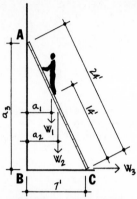

FIGURE 3-7 Man on ladder.

$$(AB)^2 + (BC)^2 = (AC)^2$$
$$AC = 24 \text{ ft} \qquad BC = 7 \text{ ft}$$
$$(AB)^2 + 7^2 = 24^2$$
$$AB = 23 \text{ ft}$$

STEP 2. Take moments about point A.

Sum of moments at A = $W_1 a_1 + W_2 a_2 - W_3 a_3 = 0$

W_1 = weight of man on ladder, 200 lb
a_1 = moment arm of man on ladder (equals horizontal length between man's foot on ladder to wall), $10/24 \times 7 = 2.92$ ft
W_2 = weight of ladder, 35 lb
a_2 = moment arm of ladder (equals horizontal length from center of ladder to wall), $12/24 \times 7 = 3.5$ ft
W_3 = horizontal force at bottom of ladder, ? lb
a_3 = moment arm of resisting force at bottom of ladder (equals vertical distance between pavement and point A), 23 ft

$$200 \times 2.92 + 35 \times 3.5 - 23W_3 = 0$$
$$W_3 = 30.7 \text{ lb}$$

STEP 3. Compute the resisting force of the meeting between ladder and pavement.

$$C \geq \frac{W}{L}$$

C = coefficient of static friction between metal and smooth concrete, 0.35

W = horizontal force at bottom of ladder, 30.7 lb

L = weight of vertical load on the pavement, $200 + 35 = 235$ lb

$$C = 0.35 \geq \frac{30.7}{235} = 0.123 \qquad \text{OK}$$

The ladder is not in danger of sliding.

Beam load distributions Formulas for common load distributions for beams and cantilevers are given in Table 3-5.

TABLE 3-5 FORMULAS FOR COMMON BEAM LOAD DISTRIBUTIONS

Load arrangement (Fig. 3-8)	Max. shear	Max. moment	Max. deflection
Simple beam, uniformly distributed load	$\dfrac{W}{2}$	$\dfrac{WL}{8}$	$\dfrac{5WL^3}{384EI}$
Simple beam, concentrated load at center	$\dfrac{P}{2}$	$\dfrac{PL}{4}$	$\dfrac{PL^3}{48EI}$
Simple beam, two equal sym. loads	P	Pa	$\dfrac{Pa(3L^2 - 4a^2)}{24EI}$

TABLE 3-5 (Continued)

Load arrangement (Fig. 3-8)		Max. shear	Max. moment	Max. deflection
Beam fixed at both ends, uniformly distributed load		$\dfrac{W}{2}$	$\dfrac{WL}{12}$	$\dfrac{WL^3}{384EI}$
Beam fixed at both ends, conc. load at center		$\dfrac{P}{2}$	$\dfrac{PL}{8}$	$\dfrac{PL^3}{192EI}$
Cantilever beam, uniformly distributed load		W	$\dfrac{WL}{2}$	$\dfrac{WL^3}{8EI}$
Cantilever beam, load increases uniformly to fixed end		W	$\dfrac{WL}{3}$	$\dfrac{WL^3}{15EI}$
Cantilever beam, concentrated load at any point		P	Pa	$\dfrac{Pa^2(3L-a)}{6EI}$

E = modulus of elasticity of beam material

I = moment of inertia of beam cross section

Example 1 What is the maximum bending moment for a 21-ft long beam that supports a uniformly distributed load (live + dead) of 120 lb/lin ft? The beam has simple end supports.

From Table 3-5, the maximum moment formula for a simple beam, uniformly distributed load is $WL/8$.

$$M = \frac{WL}{8}$$

M = maximum bending moment of load, ? in·lb
W = weight of load, 120 lb/lin ft × 21 ft = 2520 lb
L = length of beam, 21 ft = 252 in

$$M = \frac{2520 \times 252}{8}$$

$$= 79,400 \text{ in·lb}$$

Example 2   What is the maximum bending moment of a beam 24 ft long that supports a dead load of 75 lb/lin ft and a live load of two 4500-lb concentrated loads at third points on the span? The beam has simple end supports. (See Fig. 3-9).

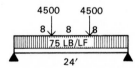

4500   4500

8   8   8

75 LB/LF

24'

FIGURE 3-9   Beam loading.

STRATEGY   Break the total load into parts whose formulas are found in Table 3-5.

M = moment for uniform load + moment for two equal
      concentrated loads

$$= \frac{WL}{8} + Pa$$

M = maximum bending moment of load, in·lb
W = weight of uniform load, 75 lb/lin ft × 24 = 1800 lb
L = length of beam, 24 ft = 288 in
P = weight of each concentrated load, 4500 lb
a = length of one-third of beam, 288 ÷ 3 = 96 in

$$M = \frac{1800 \times 288}{8} + 4500 \times 96$$
$$= 497{,}000 \text{ in} \cdot \text{lb}$$

Example 3 What is the maximum bending moment of the beam as loaded below?

STRATEGY For complicated beam loadings, draw *load, shear,* and *moment* diagrams with accompanying calculations, as shown in Fig. 3-10.

STEP 1. Find the reactions at the supports of the beam.

$$R = \frac{800 \times 6 \times 3 + 3200 \times 13}{20}$$
$$= 2800 \text{ lb}$$
$$L = 800 \times 6 + 3200 - 2800$$
$$= 5200 \text{ lb}$$

STEP 2. Draw the shear diagram.

STEP 3. Calculate the moment from the shear diagram areas.

$$M = \text{area A} + \text{area B} + \text{area C} = \text{area D}$$
$$= \frac{(5200 - 400)(6 \times 12)}{2} + 400 \times 6 \times 12 + 400$$
$$\times 7 \times 12$$
$$= 235{,}000 \text{ in} \cdot \text{lb}$$

STEP 4. Draw the moment diagram.

Column load distributions Column load arrangements are quantified in terms of the following factors:

1. Weight of the point load
2. End condition factors
3. Eccentricity of load

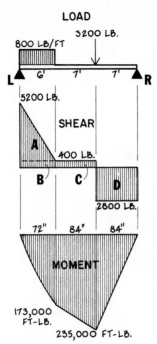

FIGURE 3-10 Beam load-shear-moment.

Example 1: Point Load The column shown in Fig. 3-11 has a load of 26.6 kips above the beams; the end reactions of the four beams connecting to the column are 7.2, 2.4, 7.2, and 3.2 kips. What is the point load on the column under the beams?

$$P = P_1 + P_2 + P_3 + P_4$$

P = total point load on the column in kips, **?** kips

P_1, P_2, P_3, P_4 = weights of each individual load on the column, as shown in Fig. 3-11.

$$P = 26.6 + 7.2 + 2.4 + 7.2 + 3.2$$
$$= 46.6 \text{ kips total load}$$

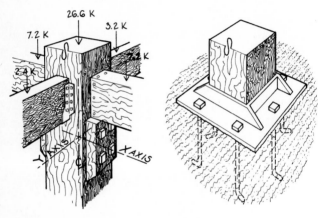

FIGURE 3-11 Column top.   **FIGURE 3-12** Column bottom.

Example 2: End Condition Factors   In the column above, the bottom is connected to its base as shown in Fig. 3-12. Considering this structural member's connections at the top and bottom, what is its end condition (K) factor?

Examine Table 3-6 to find the column's K factor. The top end condition is rotation-free, translation-fixed; the bottom end condition is rotation-free. Under these conditions, the K factor is 1.0.

TABLE 3-6 COLUMN K FACTORS

End condition	Examples (Fig. 3-13)

Rotation-
fixed

Translation-
fixed

IN X AXIS
COLUMN
END CAN'T
ROTATE
OR MOVE
LATERALLY

IN Y AXIS
COLUMN
END CAN'T
ROTATE

FOOTING

Rotation-
free

Translation-
fixed

IN X AXIS
COL. END
CAN ROTATE ABOUT CON-
NECTIONS BUT
CAN'T MOVE
LATERALLY

IN Y
AXIS
COLUMN
END CAN
ROTATE
IN SOCKET

Rotation-
fixed

Translation-
free

(FREE TRAN-
SLATION
MAY EXIST
ONLY AT
TOPS OF
COLUMNS)

IN Y AXIS
COLUMN
END CAN'T
ROTATE BUT
CAN MOVE
LATERALLY

TIER
STRUCTURE,
LIGHT CUR-
TAIN WALLS,
WIDE COL-
UMN SPCG.

Rotation-
free

Translation-
free

IN Y AXIS
COL. END
CAN ROTATE
AT JOINERY
& CAN MOVE
LATERALLY

NO DIAGONAL OR
SHEAR WALL BRAC-
ING... OTHERWISE,
END CONDITION IS
TRANSLATION FIXED

Conditions:

Column
deformations
are exaggerated
for clarity

Top:

Rotation:	fixed	free	fixed	free	fixed	free	fixed
Translation:	fixed	fixed	free	free	fixed	fixed	free

Bottom:

Rotation:	fixed	fixed	fixed	fixed	free	free	free
K factor:	0.65	0.80	1.20	2.10	0.80	1.0	2.0

Example 3: Eccentricity The column of the previous two examples is of 10 x 10 (nom.) Douglas fir. Considering the end reactions of the beams connected to it, is this column eccentrically loaded? If so, what is its equivalent load?

STEP 1. Examine the loading condition and see if any unbalanced loading exists that is not located exactly upon the central axis of the column.

Inspection reveals that the two beams in the Y axis have different end reactions and they do not rest directly upon the column's central axis. Thus the column is eccentrically loaded.

STEP 2. Determine the eccentricity of the load, as shown in Fig. 3-14.

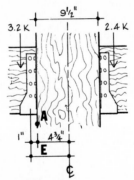

FIGURE 3-14 Eccentric load.

Load eccentricity = distance from center of column to center of
beam connection

$$E = 1 + 4.75$$
$$= 5.75 \text{ in}$$

STEP 3.   Determine the maximum equivalent load due to the eccentric loading (this occurs at A) from the following equation.

$$P_e = \frac{EPA}{S}$$

P_e = equivalent point load due to eccentricity, ? kips
E = eccentricity of load, 5.75 in
P = weight of eccentric load (3.2 kips on one side, 2.4 kips on other side), $3.2 - 2.4 = 0.8$ kips
A = area of column cross section, $9.5 \times 9.5 = 90.3$ in^2
S = section modulus of column cross section, $bd^2/6 = 9.5 \times 9.5^2/6$ $= 143$ in^3

$$P_e = \frac{5.75 \times 0.8 \times 90.3}{143} = 2.90 \text{ kips}$$

Summary   What is the total design load on the column described in the previous three examples?

$$P_d = K(P + P_e)$$

P_d = total design point load on the column, ? kips
K = column K factor due to end conditions (from "End Condition Factors" above), 1.0
P = point load on the column (from "Point Load" above), 46.6 kips
P_e = equivalent load on the column (from "Eccentricity" above), 2.90 kips

$$P_d = 1.0(46.6 + 2.9)$$
$$= 49.5 \text{ kips}$$

WOOD

Table 3-7 lists the sectional properties of standard lumber sizes.

TABLE 3-7 SECTIONAL PROPERTIES OF LUMBER SIZES

Nominal size, in	Dressed size, in	C-S area, in^2	Section modulus, in^3	Moment of inertia, in^4
2 x 2	1.5 x 1.5	2.25	0.56	0.42
2 x 3	1.5 x 2.5	3.75	1.56	1.95
2 x 4	1.5 x 3.5	5.25	3.06	5.34
2 x 6	1.5 x 5.5	8.25	7.56	20.8
2 x 8	1.5 x 7.25	10.9	13.1	47.6
2 x 10	1.5 x 9.25	13.9	21.4	98.9
2 x 12	1.5 x 11.3	16.9	31.6	178
3 x 4	2.5 x 3.5	8.75	5.10	8.93
3 x 6	2.5 x 5.5	13.8	12.6	34.7
3 x 8	2.5 x 7.25	18.1	21.9	79.4
3 x 10	2.5 x 9.25	23.1	35.7	165
3 x 12	2.5 x 11.3	28.1	52.7	297
4 x 4	3.5 x 3.5	12.3	7.15	12.5
4 x 6	3.5 x 5.5	19.3	17.6	48.5
4 x 8	3.5 x 7.25	25.4	30.7	111
4 x 10	3.5 x 9.25	32.4	50.0	231
4 x 12	3.5 x 11.3	39.4	73.8	415
6 x 2	5.5 x 1.5	8.25	2.06	1.55
6 x 4	5.5 x 3.5	19.3	11.2	19.7
6 x 6	5.5 x 5.5	30.3	27.7	76.3
6 x 8	5.5 x 7.5	41.3	51.6	194
6 x 10	5.5 x 9.5	52.3	82.7	393
6 x 12	5.5 x 11.5	63.3	121	697
8 x 2	7.25 x 1.5	10.9	2.72	2.04
8 x 4	7.25 x 3.5	25.4	14.8	25.9
8 x 6	7.5 x 5.5	41.3	37.8	104
8 x 8	7.5 x 7.5	56.3	70.3	264
8 x 10	7.5 x 9.5	71.3	113	536
8 x 12	7.5 x 11.5	86.3	165	951

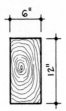

FIGURE 3-15 6 x 12 beam cross section.

Shear: Horizontal and Vertical

The beam shown in Fig. 3-15 has a maximum vertical shear of 5000 lb. What is the maximum vertical shear stress? What is the maximum horizontal shear stress?

Vertical shear:

$$v_v = \frac{V}{bd}$$

Horizontal shear:

$$v_h = \frac{3V}{2bd}$$

v_v = vertical shear unit stress, ? lb/in^2
v_h = horizontal shear unit stress, ? lb/in^2
V = total vertical shear, 5000 lb
b = width of beam, 6 in
d = depth of beam, 12 in

$$v_v = \frac{5000}{6 \times 12} = 69.4 \text{ lb/in}^2$$
$$v_h = \frac{3 \times 5000}{2 \times 6 \times 12} = 104 \text{ lb/in}^2$$

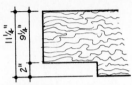

FIGURE 3-16 Notched beam.

Shear in a Notched Beam

A 4 x 12 wood beam is notched 2 in on its lower side at its end support (see Fig. 3-16). If the allowable horizontal shear stress is 85 lb/in^2, how much vertical shear can the beam resist at the notched depth?

$$v_h = \frac{3V}{2bd}$$

v_h = horizontal shear unit stress, 85 lb/in^2
V = total horizontal shear, ? lb
b = width of beam, 4 in nom. = 3.5 in
d = minimum depth of beam, 9.25 in

$$85 = \frac{3V}{2 \times 3.5 \times 9.25}$$
$$V = 1830 \text{ lb}$$

NOTE: Notches in the ends of beams should never exceed one-fourth of the beam's depth.

Bearing Area under Beam End

A 6 x 10 beam of southern pine has an end reaction of 3340 lb. What is the required length and width of the beam's bearing area?

$$A = \frac{V}{F}$$

A = minimum bearing area of beam, ? in
V = total shear end reaction, 3340 lb
F = allowable stress (use compression stress perpendicular to grain for southern pine, as listed in Table 3-8), 405 lb/in^2

TABLE 3-8 ALLOWABLE LUMBER STRESSES, lb/in²

Species, construction grade, repetitive member use, 19% moisture content	Max. bending, f_b	Vertical shear, v_v	Horizontal shear, v_h	Ten. || grain, f_t	Comp. || grain, f_c	Comp. ⊥ grain, f_p	Modulus of elasticity, E
Cedar: Northern white	675	450	65	350	625	205	600,000
Western red	875	600	75	450	850	265	900,000
Douglas fir	1200	900	95	625	1150	385	1,500,000
Hemlock, eastern	1050	600	85	525	975	365	1,000,000
Oak, white	1450	1000	200	700	900	500	1,500,000
Pine: Northern	950	900	70	475	875	280	1,100,000
Southern	1250	250	105	650	1300	405	1,500,000
Redwood	950	850	80	475	925	270	900,000
Spruce: Eastern	875	750	65	450	800	255	1,100,000
Engleman	800	700	70	400	675	195	1,000,000
Glued laminated beams, normal loads*							
Douglas fir, dry	100F	†1000	165	900	1500	385	1,700,000
Southern pine, dry	100F	1350	200	900	1500	385	1,500,000
Hem-fir, dry	100F	700	155	900	1250	245	1,600,000
Wet conditions	0.80× dry	0.88× dry	0.88× dry	0.80× dry	0.67× dry	0.73× dry	0.83× dry

*Gluelam stresses are for loads perpendicular to laminations.

†Bending unit stresses of gluelams depend on the combination symbol (such as 20F, 22F, 24F, etc.) of the structural member. *Example:* the bending unit stress of a 20F gluelam = 100 × 20 = 2000 lb/in².

$$A = \frac{3340}{405} = 8.25 \text{ in}$$

Required length and width of bearing area:

$$W = 6 \text{ in nom.} = 5.5 \text{ in}$$
$$L = \frac{\text{area}}{\text{width}} = \frac{8.25}{5.5} = 1.50 \text{ in}$$

NOTE: The minimum dimension of a bearing area involving wood is 1½ in in either direction.

Bending Moment

Example 1 A Douglas fir beam has a maximum bending moment of 66,400 in·lb. What is the most economical standard size for the beam? What is the beam's bending stress 2 in below its top?

$$M = f_b S$$

M = maximum bending moment, 66,400 in·lb
f_b = extreme fiber stress in bending (from Table 3-8, f_b for Douglas fir), 1200 lb/in^2
S = section modulus of beam, ? in^3

$$66,400 = 1200S$$
$$S = 55.3 \text{ in}^3$$

From Table 3-7, select the section moduli of standard lumber sizes that exceed 55.3 in^3 and list their cross-sectional (C-S) areas. The beam with the smallest C-S area is the most economical size.

Nominal size	Section modulus, in^3	Cross-sectional area, in^2
4 x 12	73.8	39.4+
6 x 10	82.7	52.3
8 x 8	70.3	56.3

The most economical size beam is a 4 x 12.

Now find the beam's bending stress 2 in from the top (Fig. 3-17). For 4 x 12 beam:

$$M = f_b S$$
$$66,400 = f_b \times 73.8$$
$$f_b = 900 \text{ lb/in}^2 \text{ at the top}$$

At 2 in from top of beam:

$$\frac{f_b \text{ at top}}{5.63 \text{ in}} = \frac{f_b \text{ 2 in from top}}{3.63 \text{ in}}$$

$$\frac{900}{5.63} = \frac{x}{3.63} \qquad x = 580 \text{ lb/in}^2$$

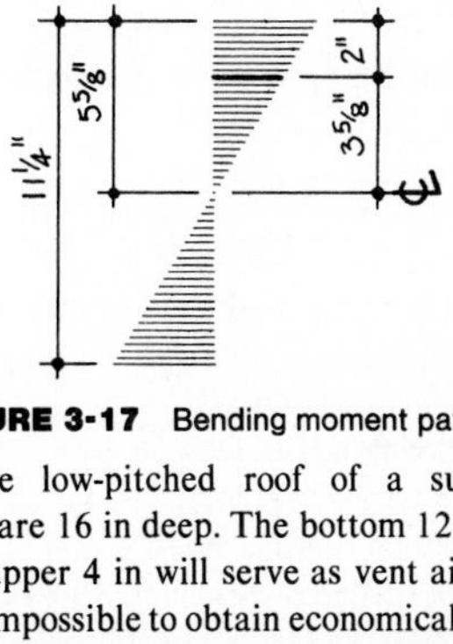

FIGURE 3-17 Bending moment pattern.

Example 2 The low-pitched roof of a superinsulated house requires rafters that are 16 in deep. The bottom 12 in will be filled with insulation, and the upper 4 in will serve as vent airspace. Because 2 x 16 wood rafters are impossible to obtain economically, composite rafters made of 2 x 3 Douglas fir and ¼-in plywood will be field assembled as shown in Fig. 3-18. If the roof loads are 30 lb/ft² live and 15 lb/ft² dead, in terms of bending moment what is the longest distance these rafters can span? For simplicity of calculation and flexibility of installation assume that $f_t = f_c = 900$ lb/in².

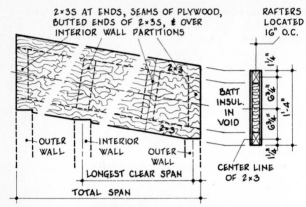

FIGURE 3-18 16-in deep rafter.

STRATEGY The moment created by the load equals the moment resisted by either the top or bottom 2 x 3.

$$M_l = M_r$$

M_l = moment created by the load; from Table 3-5, moment for uniform load on simple span = $WL/8$: W = weight of load on beam: $W = S_r(\text{LL} + \text{DL})L$:

   S_r = spacing of rafters in feet, 16 in = 1.25 ft

   LL = live load on roof, 30 lb/ft²

   DL = dead load on roof, 15 lb/ft²

    L = length of span in inches, but L in weight calculations must be expressed as feet, $L/12$ ft

   $W = 1.25(30 + 15)L/12 = 5L$ lb

    L = maximum unsupported length of rafter in inches, **?** in

$$M_l = \frac{WL}{8} = \frac{5L \times L}{8} = 0.625L^2 \text{ in·lb}$$

M_r = moment resisted by the top (or bottom) 2 x 3, $M_r = f_a Al$
  f_a = allowable bending stress for the 2 x 3, 900 lb/in^2
  A = cross-sectional area of 2 x 3, $1.5 \times 2.5 = 3.75$ in^2
  l = length of moment arm from center of 2 x 3 to center of
     built-up rafter, $16/2 - 2.5/2 = 6.75$ in

$$M_r = 900 \times 3.75 \times 6.75 = 22,800 \text{ in} \cdot \text{lb}$$

$$0.625L^2 = 22,800$$
$$L = 191 \text{ in}$$

Lateral Support

A 3 x 12 beam supporting a uniform load is 16 ft long. How much blocking or lateral bracing should the beam have?

$$L = 8D$$

L = maximum unsupported length between lateral braces, ? in
D = depth of beam, 12 in nom. = 11.3 in

$$L = 8 \times 11.3$$
$$= 90.4 \text{ in}$$

NOTE: In addition to satisfying the above criteria, lateral bracing should be installed at both ends of a beam, at locations of major point loads, and at bearing points of cantilevers.

Section Modulus

What is the section modulus about the X axis of the beam cross section shown in Fig. 3-19?

STRATEGY Find the dimensions of each piece of lumber from Table 3-7, then break up the cross-sectional area into areas whose section modulus formulas are given in Table 3-9. (See Fig. 3-20).

$$S = \frac{BD^2}{6} - \frac{bd^2}{6}$$
$$= \frac{12.3 \times 11.3^2}{6} - \frac{9.25 \times 7.75^2}{6}$$
$$= 169 \text{ in}^3$$

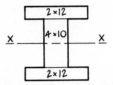

FIGURE 3-19 I beam cross section.

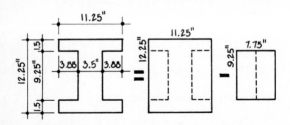

FIGURE 3-20 I beam cross-sectional breakdown.

Deflection

A 4 x 10 Douglas fir beam has a uniformly distributed load of 180 lb per running foot on a 16-ft span. If the underside of the beam is plastered and its load is 0.33 dead and 0.67 live, what is its deflection?

TABLE 3-9 PROPERTIES OF GEOMETRIC SECTIONS

Cross section (Fig. 3-21)		C-S area, A	Sec. mod., S	Moment of inertia, I
Square: Axis of moments through center		D^2	$\dfrac{D^3}{6}$	$\dfrac{D^4}{12}$
Square: Axis of moments through base		D^2	$\dfrac{D^3}{3}$	$\dfrac{D^4}{3}$
Rectangle: Axis of moments through center		BD	$\dfrac{BD^2}{6}$	$\dfrac{BD^3}{12}$
Rectangle: Axis of moments through base		BD	$\dfrac{BD^2}{6}$	$\dfrac{BD^3}{3}$
Circle: Axis of moments through center		πR^2	$\dfrac{\pi R^2}{4}$	$\dfrac{\pi R^4}{4}$

STRATEGY Calculate the maximum deflection, using the appropriate formula from Table 3-5, then compare the amount with the allowable deflection. The load arrangement is a simple beam of uniformly concentrated load.

From Table 3-5:

$$\Delta = \frac{5WL^3}{384EI}$$

Δ = deflection of beam under load conditions, ? in
W = weight of load on beam, 180 lb/lin ft $\times$ 18 ft = 3240 lb
L = length of beam in inches, 16 ft = 192 in
E = modulus of elasticity of the beam material (from Table 3-6, E for Douglas fir), 1,500,000 lb/in^2
I = moment of inertia of the beam cross section (from Table 3-7, I for a 4 x 10), 231 in^4

$$\Delta = \frac{5 \times 3240 \times 192^3}{384 \times 1,500,000 \times 231}$$
$$= 0.862 \text{ in}$$

TABLE 3-10 ALLOWABLE BEAM DEFLECTIONS

Use	Live load	Live + dead load
Industrial roof beams	$L/180$	$L/120$
All other roof and floor beams:		
With plaster or drywall ceiling	$L/360$	$L/240$
No plaster or drywall ceiling	$L/240$	$L/180$
Highway bridge stringers	$L/250$	
Railway bridge stringers	$L/350$	

Is the beam safe? Consult Table 3-10. From the table:

Use	Live load	Live + dead load
With plaster ceiling	$L/360$	$L/240$

If total Δ = 0.862 in and 0.67 of load is live, Δ due to live load = 0.862 $\times$ 0.67 = 0.578 in.

Actual $\Delta_{\text{live load}}$, 0.578 in
Allow. $\Delta_{\text{live load}}$, $L/360$ = 192/360 = 0.533 in   NG

Actual $\Delta_{\text{total load}}$, 0.862 in
Allow. $\Delta_{\text{total load}}$, $L/240 = 192/240 = 0.80$ in   NG

Glued Laminated Beams

A gluelam 5⅛ in wide made from ¾-in net laminations supports a load of 960 lb per running foot on a 28-ft span. If the beam is Douglas fir 24F, wet conditions of use, what should be its minimum depth?

STEP 1.   Solve for the maximum bending moment.

$$M = f_b S$$

M = maximum bending moment of load (load is uniformly distributed on simple span); from Table 3-5, $M = WL/8$:
   W = weight of load = $960 \times 28 = 26,900$ lb
   L = length of beam = 28 ft = 336 in

$$M = \frac{WL}{8} = \frac{26,900 \times 336}{8} = 1,130,000 \text{ in} \cdot \text{lb}$$

f_b = extreme fiber stress in bending for gluelam, Douglas fir, wet; from Table 3-8, f_b for 24F dry = 100F = $100 \times 24 = 2400$ lb/in^2; f_b for wet condition = 0.80 dry; $f_b = 0.80 \times 2400 = 1920$ lb/in^2
S = section modulus (solve for this, then in Table 3-11 find the lowest number of ¾-in plies having a larger S)

$$1,130,000 = 1920S$$
$$S = 588 \text{ in}^3$$

TABLE 3-11   SECTIONAL PROPERTIES OF GLUELAMS,* 5⅛-in width

No. 1½-in plies	No. ¾-in plies	Depth d, in	C-S area A, in²	Section modulus s, in³	Moment of inertia I, in⁴
8	16	9.00	46.1	69.2	311
	17	12.8	65.3	138	885

TABLE 3-11 (Continued)

No. 1½-in plies	No. ¾-in plies	Depth d, in	C-S area A, in²	Section modulus s, in³	Moment of inertia I, in⁴
9	18	13.5	69.2	154	1050
	19	14.3	73.0	170	1240
10	20	15.0	76.9	188	1440
	21	15.8	80.7	206	1670
11	22	16.5	84.6	226	1920
	23	17.3	88.4	240	2190
12	24	18.0	92.3	266	2490
	25	18.8	96.1	285	2820
13	26	19.5	100	309	3170
	27	20.3	104	329	3550
14	28	21.0	108	354	3960
	29	21.8	112	380	4390
15	30	22.5	115	402	4870
	31	23.3	119	429	5370
16	32	24.0	123	458	5900
	33	24.8	127	481	6480
17	34	25.5	131	511	7080
	35	26.3	135	542	7730
18	36	27.0	138	567	8400
	37	27.8	142	599	9130
19	38	28.5	146	631	9890
	39	29.3	150	665	10,700
20	40	30.0	154	692	11,500

*For complete tables of sectional properties of gluelams of all widths, see the *Timber Construction Manual*, 2d ed., pages 2-75 to 2-86.

From Table 3-11, lowest number of ¾-in plies having an S greater than 588 is 37. Thus, the depth of beam is 27.8 in.

STEP 2. Check the horizontal shear.

$$v_h \geq \frac{3V}{2bd}$$

v_h = allowable horizontal shear for gluelam, Douglas fir, wet; from Table 3-8, v_h for Douglas fir dry = 165 lb/in²; v_h for wet conditions = 0.88 dry; v_h = 0.88 × 165 = 145 lb/in²

V = maximum shear; from Table 3-5, maximum shear for simple beam with uniform load = $W/2$; in calculating W, neglect all beam loads within a distance d (depth of beam) from end of span; thus:

$$V = \frac{W}{2} - \frac{w \times d}{12}$$

W = weight of load on beam, from step 1, 26,900 lb
w = load per linear foot, 960 lb/lin ft
d = depth of beam, from step 1, 27.8 in

$$V = \frac{26,900}{2} - \frac{960 \times 27.8}{12} = 11,200 \text{ lb}$$

b = width of beam, 5⅛ in = 5.13 in
d = depth of beam, 27.8 in

$$145 \geq \frac{3 \times 11,200}{2 \times 5.13 \times 27.8} = 118 \text{ lb/in}^2 \qquad \text{OK}$$

STEP 3. Check for deflection.
Actual Δ. From Table 3-5:

$$\Delta = \frac{5WL^3}{384EI}$$

W = weight of load on beam from step 1, 26,900 lb
L = length of beam from step 1, 336 in
E = modulus of elasticity for Douglas fir gluelam, wet conditions; from Table 3-8, E for Douglas fir dry = 1,700,000 lb/in^2; = 1,700,000 × 0.83 = 1,410,000 lb/in^2
I = moment of inertia for 5.13 E for wet conditions = 0.83 dry; E × 27.8 in gluelam; from Table 3-11, I = 9.30 in^4

$$\Delta = \frac{5 \times 26,900 \times 336^3}{384 \times 1,410,000 \times 9130} = 1.03 \text{ in}$$

Allowable Δ. From Table 3-10:

$$\Delta \text{ due to live + dead load} = \frac{L}{180}$$

$$1.03 \geq \frac{336}{180} = 1.87 \text{ in} \qquad \text{OK}$$

Warren Truss Design

A Warren truss is loaded as shown in Fig. 3-22. What is the axial force in member A?

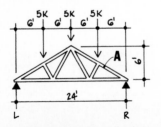

FIGURE 3-22 Warren truss.

STEP 1. Determine the reactions at the truss ends.

Forces *up* = forces *down*
$$R + L = 15 \text{ kips}$$

By symmetry, $R = L$. Thus, $R = L = 7.5$ kips.

STEP 2. Isolate the right truss end with a free body diagram (Fig. 3-23).

Pitch of A = 6 in 12

$$\tan \phi = 6/12 = 0.5$$
$$\phi = 26.6°$$

Vertical forces at truss end:

$$7.5 = A \sin \phi = A \sin 26.6° = A \times 0.447$$
$$A = 16.8 \text{ kips}$$

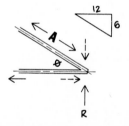

FIGURE 3-23 Warren truss detail.

NOTE: At any point in the truss, the sum of vertical forces = 0, sum of horizontal forces = 0, sum of moments about any point = 0, and a free body diagram may be drawn through any part.

Pratt Truss Design

A Pratt truss is loaded as shown in Fig. 3-24. What is the axial force in member J? What is unusual about member K?

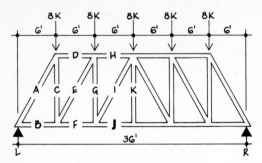

FIGURE 3-24   Pratt truss 1.

STEP 1.   Determine the reactions at the truss ends.
Sum of vertical forces:

$$R + L = 8 + 8 + 8 + 8 + 8 = 40 \text{ kips}$$
$$R = L \text{ by symmetry}$$
$$R = L = 20 \text{ kips each}$$

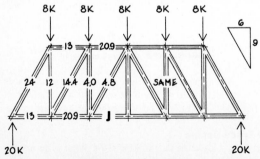

FIGURE 3-25   Pratt truss 2.

STEP 2.   Find the axial forces in struts A through I in alphabetical order by taking the sum of verticals and the sum of horizontals about the panel points. After finding the value of member I, the truss loading will look like Fig. 3-25.

STEP 3.   Compute the axial force in strut J. (See Fig. 3-26.)

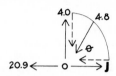

FIGURE 3-26   Pratt truss detail.

Sum of horizontal forces:

$$J = 20.9 + 4.8 \cos \phi$$

ϕ has a 9 in 6 pitch. Thus:

$$\phi = \tan^{-1} 9/6 = 56.3°$$
$$J = 20.9 + 4.8 \cos 56.3° = 23.6 \text{ kips}$$

What is unusual about K? K = 0.

Truss Chord Design

Design truss member J above, using standard Douglas fir lumber.

$$P = f_t A$$

P = point load on the truss member, 23.5 kips × 1000 = 23,500 lb
f_t = allowable stress in truss member (from Table 3-8, allowable stress in tension parallel to grain for Douglas fir), 625 lb/in²
A = cross-sectional area of truss member, ? in²

$$23,500 = 625A$$
$$A = 37.6 \text{ in}^2$$

From Table 3-7, select the smallest lumber size whose A exceeds 37.6 in^2.

$$\text{C-S area of 4 x 12} = 39.4 \text{ in}^2 +$$
$$\text{C-S area of 6 x 8} = 41.3 \text{ in}^2$$

The 4 x 12 is the economical section.

Column: Concentric Load

A column of southern pine supports a load of 43.2 kips. The column is 19 ft 4 in tall, its load is axial, and its end condition factor is 1.0. If standard timber is used, what is the column's economical cross section?

STRATEGY Square columns are the most efficient. Try a square column cross section and use its thickness (depth) to find the column's trial width. Then consult the table below.

If the trial width is:	*Do the following:*
1. Much less than the trial depth	Try a smaller square C-S
2. Slightly less than the trial depth	Trial C-S is OK
3. Slightly more than the trial depth	Use d = trial d and w = next larger nominal size
4. Much more than the trial depth	Try a larger square C-S

STEP 1. Pick a square column thickness. Make sure that $L/D \leq$ 50.

Try $D = 8$ in nom. (7.5 in).

$$\frac{L}{D} = \frac{19 \text{ ft 4 in}}{7.5 \text{ in}} = 30.9 < 50 \qquad \text{OK}$$

STEP 2. Find the allowable compression unit stress from the following formula.

$$0.3E = f_c \left(\frac{L}{D}\right)^2$$

E = modulus of elasticity (as listed in Table 3-8, E for southern pine), 1,500,000 lb/in^2

f_c = allowable compression unit stress, ? lb/in^2

L = unbraced length of column, 232 in

D = minimum dimension of column cross section, 7.5 in

$$0.3 \times 1,500,000 = f_c \left(\frac{232}{7.5}\right)^2$$
$$f_c = 471 \text{ lb/in}^2$$

STEP 3. Find the trial width from the following formula.

$$P = f_c DW$$

P = design point load (eccentricity = 0, K = 1), 43.2 $\times$ 1000 = 43,200 lb

f_c = allowable compression unit stress, 471 lb/in^2

D = minimum dimension of column cross section, 7.5 in

W = width of column cross section, ? in

$$43,200 = 471 \times 7.5D$$
$$D = 12.2 \text{ in}$$

Use 14 in nom. Column size is 8 x 14 in nom. This is much more than the trial depth (7.5 in) and thus is NG.

STEP 4.   Because the first column size was too small, go back to step 1 and try a larger square column size.

Try $D = 10$ in nom (9.5 in).

$$0.3E = f_c \left(\frac{L}{D} \right)^2$$

$E = 1,500,000$, $L = 232$ in, $D = 9.5$ in, $f_c = ?$

$$0.3 \times 1,500,000 = f_c \left(\frac{232}{9.5} \right)^2$$
$$f_c = 755 \text{ lb/in}^2$$

$$P = f_c DW$$

$P = 43,200$ lb, $f_c = 755$ lb/in^2, $D = 9.5$ in, $W = ?$

$$43,200 = 755 \times 9.5W$$
$$W = 6.02 \text{ in}$$

W is considerably less than D, and you may at first think to try a smaller square cross section. But your first trial was the next smallest nominal size, and thus in this case, W should equal D. So a 10×10 nom. is the economical section.

Column: Eccentric Load

What are the fiber stresses at points A and B in the eccentrically loaded column shown in Fig. 3-27?

STRATEGY   This eccentrically loaded column creates a combined compression and bending stress at A and B, the points of extreme fiber stress on the sides of the column. At A, the compressive face, the total stress $= P/A + M/S$. At B, the tensile face, the total stress $= P/A - M/S$.

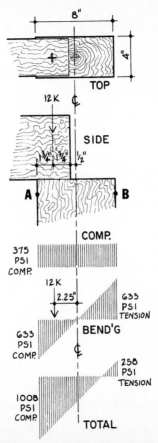

FIGURE 3-27 Beam-to-column eccentric load.

Compressive stress:

$$f_c = \frac{P}{A}$$

P = point load, 12 kips = 12×1000 = 12,000 lb
A = cross-section area, 8×4 = 32 in^2

$$f_c = \frac{12,000}{32} = 375 \text{ lb/in}^2$$

Bending stress:

$$M = f_b S$$

M = moment created by the load, $M = Pe$
  P = 12,000 lb
  e = eccentricity of load, $1.75 + 0.5 = 2.25$ in
  $$M = 12,000 \times 2.25 = 27,000 \text{ in·lb}$$
f_b = extreme fiber stress due to bending, ? lb/in^2
S = section modulus of cross section, $S = bd^2/6$
  b = 4 in
  d = 8 in

$$S = \frac{4 \times 8^2}{6} = 42.7 \text{ in}^3$$

$$27,000 = f_b \times 42.7$$
$$f_b = 633 \text{ lb/in}^2$$

At point A, total stress = $375 + 633 = 1008$ lb/in^2 compression
At point B, total stress = $375 - 633 = 258$ lb/in^2 tension

Combined Compression and Bending

A redwood post 12 ft 8 in tall supports an axial load of 14,500 lb and a uniform load of 200 lb/lin ft against its side. What is the smallest standard timber size that can be used?

STEP 1.   Try a square column size in the formula below. L/D must ≤ 50. Try an 8 x 8:

$$L = 12 \text{ ft } 8 \text{ in} = 12 \times 12 + 8 = 152 \text{ in}$$
$$D = 8 \text{ in nom.} = 7.5 \text{ in}$$
$$\frac{L}{D} = \frac{152}{7.5} = 20.3 \leq 50 \qquad \text{OK}$$

$$\frac{PL^2}{0.3AED^2} + \frac{M}{Sf_b} = 1$$

P = point load of the column, 14,500 lb

L = unbraced length of column, 12 ft 8 in = 152 in

A = area of column cross section, $7.5 \times 7.5 = 56.3 \text{ in}^2$

E = modulus of elasticity (from Table 3-8, E for redwood), 900,000 lb/in^2

D = minimum dimension of the column cross section, 7.5 in

M = bending moment for uniform load against the column; From Table 3-5, M for this condition = $WL/8$:

W = uniform load against column, 200 lb/lin ft $\times$ 12.7 ft = 2540 lb

L = length of load (in this case, height of column), 12 ft 8 in high = 152 in

$$\frac{WL}{8} = \frac{2540 \times 152}{8} = 48,300 \text{ in} \cdot \text{lb}$$

S = section modulus of column cross section (from Table 3-7, S for an 8 x 8 nom. cross section), 70.3 in^3

f_b = allowable bending stress for redwood (from Table 3-8, f_b), 950 lb/in^2

$$\frac{14,500 \times 152^2}{0.3 \times 56.3 \times 900,000 \times 7.5^2} + \frac{48,300}{70.3 \times 950} \leq 1$$
$$0.39 + 0.72 = 1.11 \qquad \text{Not} \leq 1$$

8 x 8 is NG

STEP 2.   Because an 8 x 8 is almost big enough, try an 8 x 10 in the same formula. With an 8 x 10:

$$A = 7.5 \times 9.5 = 71.3 \text{ in}^2$$
$$S = \frac{BD^2}{6} = \frac{9.5 \times 7.5^2}{6} = 89.1 \text{ in}^3$$

All other values are the same.

$$\frac{14,500 \times 152^2}{0.3 \times 71.3 \times 900,000 \times 7.5^2} + \frac{48,100}{89.1 \times 950} \le 1$$
$$0.31 + 0.57 = 0.88 \le 1$$

8 x 10 is OK.

Connections: Single and Double Shear

Wood corbels are to be installed as shown in Fig. 3-28 to transfer the weight of the beams to the column. The end reactions of the four beams

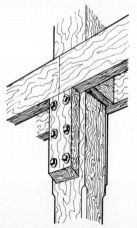

FIGURE 3-28   Wood corbel connection.

are 7.4 kips each, the post is an 8 x 8, the corbels are made of 3 x 6s, and all wood is Douglas fir. If six metal bolts hold the corbels to the column, what are their diameters? What is the length of the corbels?

STEP 1.   Find the load on each bolt.

$$P = NB$$

P = total point load on corbels (4 beams $\times$ 7.4 kips each), 29.6 kips
N = number of volts, 6
B = load on each bolt, ? kips

$$29.6 = 6B$$
$$B = 4.93 \text{ kips or } 4930 \text{ lb each}$$

TABLE 3-12   HOLDING POWER OF BOLTS

Length of bolt in main wood member, in			Diameter of bolt, in					
			⅜	½	⅝	¾	⅞	1
1½	Single shear	⊥ grain	330	430	490	540	600	650
		‖ grain	370	650	1000	1350	1650	1920
	Double shear	⊥ grain	370	430	490	540	600	650
		‖ grain	670	960	1210	1460	1700	1940
2½	Single shear	⊥ grain		480	710	890	990	1080
		‖ grain		650	1020	1470	1990	2590
	Double shear	⊥ grain	620	720	810	900	990	1080
		‖ grain	730	1290	1870	2370	2810	3230
3½	Single shear	⊥ grain		660	930	1220	1470	1650
		‖ grain		1020	1470	2000	2610	3300
	Double shear	⊥ grain	980	1130	1260	1390	1520	1650	1⅛
		‖ grain	1300	2030	2870	3670	4380	5040

TABLE 3-12 HOLDING POWER OF BOLTS *(Continued)*

Length of bolt in main wood member, in			*Diameter of bolt, in*						
			½	⅝	¾	⅞	1	1⅛	
5½	Single shear	⊥ grain				1060	1410	1800	
		∥ grain				1990	2610	3300	
	Double shear	⊥ grain	930	1410	1880	2180	2380	2600	
		∥ grain	1300	2040	2930	4000	5200	6540	
			½						
7½	Single shear	⊥ grain						1980	
		∥ grain						4080	
	Double shear	⊥ grain	1260	1820	2430	3030	3500	3800	1¼
		∥ grain	2040	2930	3990	5210	6610	8150	
			⅝						
9½	Single shear	⊥ grain						2650	
		∥ grain						5870	
	Double shear	⊥ grain	1640	2270	2960	3710	4450	5530	1½
		∥ grain	2930	4000	5210	6600	8150	11750	
11½	Single shear	⊥ grain							
		∥ grain							
	Double shear	⊥ grain		2050	2770	3540	4360	6150	
		∥ grain		4000	5210	6600	8150	11740	
			¾	⅞	1	1⅛	1¼	1½	

STEP 2. Select the proper bolt diameter from Table 3-12. First, determine the following:

Length of bolt in main wood member. 8 x 8 column indicates the length of bolt through it is *7.5 in.*

Single or double shear? Shear planes on both sides of column indicate *double shear*.

Load perpendicular or parallel to grain? Wood grains reveal that the load is *parallel to grain*.

In Table 3-12, begin on the far left with bolt length (7½ in), move to the right through "double shear" and "∥ to grain" to the first number that is larger than the bolt load of 4930 lb, then read up to obtain the bolt size. It is 1-in diameter.

STEP 3.   Determine the length of the corbel. The beam loads are transferred to the bolts through the shear planes in the corbels as shown in Figure 3-29. This stress is shear parallel to grain.

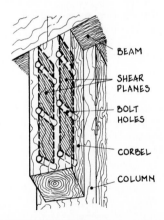

FIGURE 3-29   Shear planes in wood corbel.

$$P = v_h A$$

P = weight of load on each corbel, $7.4 \times 2 = 14.8$ kips $= 14,800$ lb

v_h = allowable shear parallel to grain for Douglas fir (from Table 3-8, v_h for Douglas fir), 95 lb/in²
A = cross-sectional area of shear planes, $A = TLN$:
  T = thickness of corbel, 3 in nom. = 2.5 in
  L = minimum length of shear plane, ? in
  N = number of shear planes in each corbel, 2
  $A = 2.5 \times L \times 2 = 5L$

$$14,800 = 95 \times 5L$$
$$L = 31.2 \text{ in}$$

Nail Strength and Penetration

Example 1 2 x 8 deck joists are nailed to a 3 x 12 girder as shown in Fig. 3-30. Assume that the 2 x 4 sill plate supports the total end reaction of the 2 x 8s, which is 420 lb at 16 in o.c. If 16d common nails hold the sill plate to the girder, what is their spacing? All lumber is Douglas fir.

$$L = NS$$

L = total load transferred through the nails, 420 lb per 16 in
N = number of nails required to transfer the load, ?
S = safe lateral strength of each nail (as listed in Table 3-13, S for 16d nails in Douglas fir), 107 lb each.

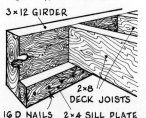

FIGURE 3-30 2 x 8 joists into 3 x 12 girder.

TABLE 3-13 SAFE LATERAL STRENGTH OF COMMON NAILS*

				Size of nail				
Length of nail	6d	8d	10d	12d	16d	20d	30d	J.H.†
Penetration into parent wood, in	2	2½	3	3¼	3½	4	4½	2¾
Cedar: Northern, white	20	25	29	30	34	44	49	67
Western red	28	35	41	43	48	63	69	92
Douglas fir	63	78	92	95	107	139	154	211
Hemlock, eastern	48	59	70	72	81	106	117	160
Oak, white	124	154	181	187	211	274	303	414
Pine: Northern	48	59	70	72	81	106	117	160
Southern	63	78	92	95	107	139	154	211
Redwood	35	44	52	53	60	78	86	118
Spruce: Eastern	40	49	58	60	67	87	97	133
Engelman	27	33	39	40	45	58	65	89

Clinched nails (min. 3 nail diameters): 2.0 x above

Nails driven into end grain: 0.67 x above

Nails through metal side plates: 1.25 x above

Toenails (45° to 55° angle): 0.83 × above

Safe withdrawal loads (tension parallel to shank) 0.33 × above per inch penetration.

*Measured in pounds per nail driven normal to side grain of wood.

†J.H. = heavy-duty joist hanger nails

$$420 = N \times 107$$
$$N = 3.92 \qquad \text{minimum}$$

Use four nails every 16 in.

Example 2   What is the safe withdrawal strength of four 16d common nails toe-nailed through a top plate into the top of a 4 × 4 Western red cedar post, as shown in Fig. 3-31?

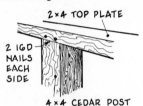

FIGURE 3-31   Top plate on post.

From Table 3-13:

16d common nail into Western red cedar, 48 lb per inch depth
Nail driven into end grain, 0.67 × above
Toenails, 0.83 × above
Safe withdrawal load, 0.33 × above per inch penetration
Assume 2 in penetration, 2.0 × above
Four nails in connection, 4.0 × above

Safe withdrawal strength of connection:

$$48 \times 0.67 \times 0.83 \times 0.33 \times 2 \times 4 = 70 \text{ lb}$$

Withdrawal Strength of Threaded Connectors

Example 1   A 40-in balcony railing has 2 x 3 wood balusters at 12-in centers. The bottoms of the balusters are held to the edge of the balcony by lag bolts passing through the 2 x 4s and anchoring in a 3 x 12 Douglas fir girder. If the lag bolts are located 1½ in from the top of the 3 x 12 and the top of the railing is designed to withstand a horizontal force of 125 lb/lin ft, what diameter should be the lag bolts? See Fig. 3-32.

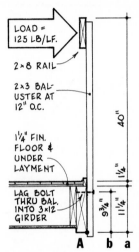

LOAD = 125 LB/LF.

2 × 8 RAIL

2 × 3 BALUSTER AT 12" O.C.

1¼" FIN. FLOOR & UNDER LAYMENT

LAG BOLT THRU BAL. INTO 3 × 12 GIRDER

40"

1¼"

9¾" 1¼"

A   b   a

FIGURE 3-32 Balcony cross section.

STEP 1.   Take moments about point A.

$$Pa = Rb$$

P = horizontal force at top of rails, 125 lb/lin ft
a = moment arm length from point A to top of railing, 40 + 1.25 + 11.25 = 52.5 in
R = resisting force of lag bolt in 3 x 12 girder, **?** lb per baluster

TABLE 3-14 WITHDRAWAL STRENGTH OF THREADED CONNECTORS

		Withdrawal strength, lb/in penetration*			
		Group A	Group B	Group C	Group D
Connector	Size	specific gravity = 0.34	specific gravity = 0.42	specific gravity = 0.48	specific gravity = 0.60
Wood screws	no. 6	45	68	90	139
	7	50	76	98	153
	8	55	82	108	167
	9	59	89	115	178
	10	63	96	125	194
	12	72	109	141	220
	14	81	124	160	248
	16	90	137	178	277
Lag bolts:					
diameters, in	¼	126	172	210	295
	5/16	150	206	251	350
	⅜	173	238	289	400
	7/16	193	265	325	450
	½	215	295	360	500
	9/16	235	320	391	545
	⅝	253	346	421	589
	¾	288	395	481	671
	⅞	329	450	550	770
	1	359	492	601	840
	1⅛	392	538	656	920
	1¼	425	582	710	995

Wood group A: cedar, redwood, white pine, spruce
Wood group B: hemlock, jack pine, hem-fir
Wood group C: Douglas fir, southern pine, tupelo
Wood group D: oak, ash, birch, maple
*Values are for connectors installed $\perp$ side grain. For end grain connections use $0.67 \times$ value.

b = moment arm length from A to center of lag bolt, $11.25 - 1.5 =$ 9.75 in

$$125 \times 52.5 = R \times 9.75$$
$$R = 673 \text{ lb per baluster}$$

STEP 2.   Calculate the diameter of the bolt.

$$RS = 1800G^{1.5}D^{0.75}P$$

R = resisting force of lag bolt in 3 x 12 wood girder, 673 lb
S = spacing of lag bolts in girder (number per linear foot; balusters are 1 ft o.c.), 1 per linear foot
G = specific gravity of wood species in which the lag bolt is anchored (as listed in Table 3-14, species is Douglas fir, which is a type B wood), 0.48
D = diameter of lag bolt, ? in
P = penetration of lag bolt in 3 x 12 girder (girder is 2½ in thick; assume maximum penetration), 2.5 in

$$673 \times 1 = 1800 \times 0.48^{1.5} \times D^{0.75} \times 2.5$$
$$D^{0.75} = 0.449$$
$$D = 0.344 \text{ in} \qquad \text{minimum}$$

Use ⅜-in diameter bolts.

Example 2   A 3¼-in long screw hook (whose shank has the same diameter and threads as a no. 10 wood screw) is screwed into a 2 x 10 red cedar greenhouse rafter where it will hold a potted plant. What is the maximum weight of the plant that can hang on the screw hook?

Consult Table 3-14. Find "cedar" in wood group A. In column A, read down to the number alongside "wood screw, no. 10." The answer is 63 lb. (See Fig. 3-33.)

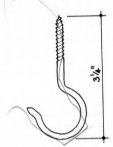

3¼"

FIGURE 3-33   Screw hook.

Hinge Size

The entrance door to a community medical facility is 1¾ in thick, 42 in wide, and 8 ft tall. If the door is solid core, how many hinges should it have and what size should they be?

Number of hinges:

Hollow-core doors, 2 hinges
Solid-core doors, 3 hinges

Hinge size:

$$S = 0.02(HWT)^{0.63}$$

S = size of hinge measured along length of pin, ? in
H = height of door in inches, 8 ft = 96 in
W = width of door in inches, 42 in
T = thickness of door in inches, 1¾ in = 1.75 in

$$S = 0.02(96 \times 42 \times 1.75)^{0.63}$$
$$= 5.32 \text{ in}$$

Use 5½-in standard or 5-in heavy duty hinges.

STEEL

Table 3-15 lists the allowable stresses for various kinds of steel, while Table 3-16 lists the dimensions and proportions of structural steel sections.

TABLE 3-15 ALLOWABLE STEEL STRESSES, kips/in²

		Shapes, plates, bars			Bolts, rivets	
		Tens. comp. bending			Tens. &	
		Standard*	Compact*	Shear	comp.	Shear
Steel type	f_y	$f_t\,f_c\,f_b$	$f_t\,f_c\,f_b$	f_v	$f_t\,f_c$	f_v
A36	36.0	22.0	24.0	14.5	19.1	9.9
A529	42.0	25.2	27.5	17.0	20.0	13.2
A441	40.0	24.0	26.4	16.0	—	—
	46.0	27.6	30.4	18.4	—	—
A572	42.0	25.2	27.5	17.0	19.8	13.2
	50.0	30.0	33.0	20.0	21.5	11.1
A242	50.0	30.0	33.0	20.0	—	—
A588	50.0	30.0	33.0	20.0	23.1	15.4
A307	—	—	—	—	20.0	10.0
A325	81.0	—	—	—	44.0	21.0
A449	81.0	—	—	—	34.7	23.1
A490	—	—	—	—	54.0	28.0
A502	88.0	—	—	—	29.0	22.0
A615	60.0	36.0	39.6	24.0	—	—

*Unless otherwise noted, steel shapes are compact.

TABLE 3-16 DIMENSIONS AND PROPERTIES OF STEEL SECTIONS*

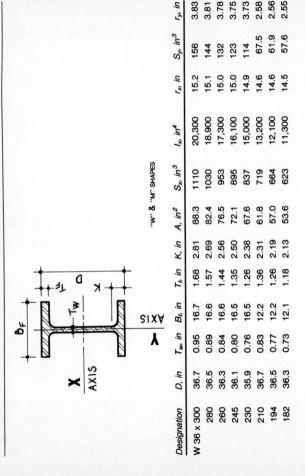

"W" & "M" SHAPES

Designation	D, in	T_w, in	B_h, in	T_h, in	K, in	A, in^2	S_x, in^3	I_x, in^4	r_x, in	S_y, in^3	r_y, in
W 36 x 300	36.7	0.95	16.7	1.68	2.81	88.3	1110	20,300	15.2	156	3.83
280	36.5	0.89	16.6	1.57	2.69	82.4	1030	18,900	15.1	144	3.81
260	36.3	0.84	16.6	1.44	2.56	76.5	953	17,300	15.0	132	3.78
245	36.1	0.80	16.5	1.35	2.50	72.1	895	16,100	15.0	123	3.75
230	35.9	0.76	16.5	1.26	2.38	67.6	837	15,000	14.9	114	3.73
210	36.7	0.83	12.2	1.36	2.31	61.8	719	13,200	14.6	67.5	2.58
194	36.5	0.77	12.2	1.26	2.19	57.0	664	12,100	14.6	61.9	2.56
182	36.3	0.73	12.1	1.18	2.13	53.6	623	11,300	14.5	57.6	2.55

170	36.2	0.68	12.0	1.10	2.00	50.0	580	10,500	14.5	53.2	2.53
160	36.0	0.65	12.0	1.02	1.94	47.0	542	9750	14.4	49.1	2.50
150	35.9	0.63	12.0	0.94	1.88	44.2	504	9040	14.3	45.1	2.47
135	35.5	0.60	12.0	0.79	1.69	39.7	439	7800	14.0	37.7	2.38
W 33 x 241	34.2	0.83	15.9	1.40	2.19	70.9	829	14,200	14.1	118	3.63
221	33.9	0.78	15.8	1.28	2.06	65.0	757	12,800	14.1	106	3.58
201	33.7	0.72	15.7	1.15	1.94	59.1	684	11,500	14.0	95.2	3.56
152	33.5	0.64	11.6	1.06	1.88	44.7	487	8160	13.5	47.2	2.47
141	33.3	0.61	11.5	0.96	1.75	41.6	448	7450	13.4	42.7	2.43
130	33.1	0.58	11.5	0.86	1.69	38.3	406	6710	13.2	37.9	2.39
118	32.9	0.55	11.5	0.74	1.56	34.7	359	5900	13.0	32.6	2.32
W 30 x 211	30.9	0.78	15.1	1.32	2.13	62.0	663	10,300	12.9	100	3.49
191	30.7	0.71	15.0	1.19	1.94	56.1	598	9170	12.8	89.5	3.46
173	30.4	0.66	15.0	1.07	1.88	50.8	539	8200	12.7	79.8	3.43
132	30.3	0.62	10.5	1.00	1.75	38.9	380	5770	12.2	37.2	2.25
124	30.2	0.59	10.5	0.93	1.69	36.5	355	5360	12.1	34.4	2.23
116	30.0	0.57	10.5	0.85	1.63	34.2	329	4930	12.0	31.3	2.19
108	29.8	0.55	10.5	0.76	1.56	31.7	299	4470	11.9	27.9	2.15
99	29.7	0.52	10.5	0.67	1.44	29.1	269	3990	11.7	24.5	2.10
W 27 x 178	27.8	0.73	14.1	1.19	1.88	52.3	502	6990	11.6	78.8	3.26
161	27.6	0.66	14.0	1.08	1.81	47.4	455	6280	11.5	70.9	3.24

TABLE 3-16 DIMENSIONS AND PROPERTIES OF STEEL SECTIONS* *(Continued)*

"W" & "M" SHAPES

Designation	D, in	T_w, in	B_f, in	T_f, in	K, in	A, in^2	S_x, in^3	I_x, in^4	r_x, in	S_y, in^3	r_y, in
146	27.4	0.61	14.0	0.98	1.69	42.9	411	5630	11.4	63.5	3.21
114	27.3	0.57	10.1	0.93	1.63	33.5	299	4090	11.0	31.5	2.18
102	27.1	0.52	10.0	0.83	1.56	30.0	267	3620	11.0	27.8	2.15
94	26.9	0.49	9.99	0.75	1.44	27.7	243	3270	10.9	24.8	2.12
84	26.7	0.46	9.96	0.64	1.38	24.8	213	2850	10.7	21.2	2.07
W 24 × 162	25.0	0.71	13.0	1.22	2.00	47.7	414	5170	10.4	68.4	3.05
146	24.7	0.65	12.9	1.09	1.88	43.0	371	4580	10.3	60.5	3.01
131	24.5	0.61	12.9	0.96	1.75	38.5	329	4020	10.2	53.0	2.97
117	24.3	0.55	12.8	0.85	1.63	34.4	291	3540	10.1	46.5	2.94
104	24.1	0.50	12.8	0.75	1.50	30.6	258	3100	10.1	40.7	2.91
94	24.3	0.52	9.07	0.88	1.63	27.7	222	2700	9.87	24.0	1.98
84	24.1	0.47	9.02	0.77	1.56	24.7	196	2370	9.79	20.9	1.95
76	23.9	0.44	8.99	0.68	1.44	22.4	176	2100	9.69	18.4	1.92
68	23.7	0.42	8.97	0.59	1.38	20.1	154	1830	9.55	15.7	1.87
62	23.7	0.43	7.04	0.59	1.38	18.2	131	1550	9.23	9.80	1.38
55	23.6	0.40	7.00	0.51	1.31	16.2	114	1350	9.11	8.30	1.34
W 21 × 147	22.1	0.72	12.5	1.15	1.88	43.2	329	3630	9.17	60.1	2.95
132	21.8	0.65	12.4	1.04	1.81	38.8	295	3220	9.12	53.5	2.93

122	21.7	0.60	12.4	0.96	1.69	35.9	273	2960	9.09	49.2	2.92
111	21.5	0.55	12.3	0.88	1.62	32.7	249	2670	9.05	44.5	2.90
101	21.4	0.50	12.3	0.80	1.56	29.8	227	2420	9.02	40.3	2.89
93	21.6	0.58	8.42	0.93	1.69	27.3	192	2070	8.70	22.1	1.84
83	21.4	0.52	8.36	0.84	1.56	24.3	171	1830	8.67	19.5	1.83
73	21.2	0.46	8.30	0.74	1.50	21.5	151	1600	8.64	17.0	1.81
68	21.1	0.43	8.27	0.69	1.55	20.0	140	1480	8.60	15.7	1.80
62	21.0	0.40	8.24	0.62	1.38	18.3	127	1330	8.54	13.9	1.77
57	21.1	0.41	6.56	0.65	1.38	16.7	111	1170	8.36	9.35	1.35
50	20.8	0.38	6.53	0.54	1.31	14.7	94.5	984	8.18	7.64	1.30
44	20.7	0.35	6.50	0.45	1.19	13.0	81.6	843	8.06	6.36	1.26
W 18 x 119	19.0	0.66	11.3	1.06	1.75	35.1	231	2190	7.90	44.9	2.69
106	18.7	0.59	11.2	0.94	1.63	31.1	204	1910	7.84	39.4	2.66
97	18.6	0.54	11.1	0.87	1.56	28.5	188	1750	7.82	36.1	2.65
86	18.4	0.48	11.1	0.77	1.44	25.3	166	1530	7.77	31.6	2.63
76	18.2	0.43	11.0	0.68	1.38	22.3	146	1330	7.73	27.6	2.61
71	18.5	0.50	7.64	0.81	1.50	20.8	127	1170	7.50	15.8	1.70
65	18.4	0.45	7.59	0.75	1.44	19.1	117	1070	7.49	14.4	1.69
60	18.2	0.42	7.56	0.70	1.38	17.6	108	984	7.47	13.3	1.69
55	18.1	0.39	7.53	0.63	1.31	16.2	98.3	890	7.41	11.9	1.67
50	18.0	0.36	7.50	0.57	1.25	14.7	88.9	800	7.38	10.7	1.65
46	18.1	0.36	6.06	0.61	1.25	13.5	78.8	712	7.25	7.43	1.29
40	17.9	0.32	6.02	0.53	1.18	11.8	68.4	612	7.21	6.35	1.27

TABLE 3-16 DIMENSIONS AND PROPERTIES OF STEEL SECTIONS* *(Continued)*

"W" & "M" SHAPES

Designation		D, in	T_w, in	B_h, in	T_b, in	K, in	A, in²	S_x, in³	I_x, in⁴	r_x, in	S_y, in³	r_y, in
	35	17.7	0.30	6.00	0.43	1.13	10.3	57.6	510	7.04	5.12	1.22
	100	17.0	0.59	10.4	0.99	1.69	29.4	175	1490	7.10	35.7	2.51
	89	16.8	0.53	10.4	0.88	1.56	26.2	155	1300	7.05	31.4	2.49
	77	16.5	0.46	10.3	0.76	1.44	22.6	134	1100	7.00	26.9	2.47
	67	16.3	0.40	10.2	0.67	1.38	19.7	117	954	6.96	23.2	2.46
	57	16.4	0.43	7.12	0.72	1.38	16.8	92.2	758	6.72	12.1	1.60
W 16 x	50	16.3	0.38	7.07	0.63	1.31	14.7	81.0	659	6.68	10.5	1.59
	45	16.0	0.35	7.04	0.57	1.25	13.3	72.7	586	6.65	9.34	1.57
	40	16.0	0.31	7.00	0.51	1.19	11.8	64.7	518	6.63	8.25	1.57
	36	15.9	0.30	6.99	0.43	1.13	10.6	56.5	448	6.51	7.00	1.52
	31	15.9	0.28	5.53	0.44	1.13	9.12	47.2	375	6.41	4.49	1.17
	26	15.7	0.25	5.50	0.35	1.06	7.68	38.4	301	6.26	3.49	1.12
W 14 x	730	22.4	3.07	17.9	4.91	5.56	215	1280	14,300	8.17	527	4.69
	550	20.2	2.38	17.2	3.82	4.50	162	931	9430	7.63	378	4.49
	426	18.7	1.88	16.7	3.04	3.69	125	707	6600	7.26	283	4.34
	283	16.7	1.29	16.1	2.07	2.75	83.3	459	3840	6.79	179	4.17
	132	14.7	0.65	14.7	1.03	1.69	38.8	209	1530	6.28	74.5	3.76
	120	14.5	0.59	14.7	0.94	1.63	35.3	190	1380	6.24	67.5	3.74

109	14.3	0.53	14.6	0.86	1.56	32.0	173	1240	6.22	61.2	3.73
99	14.2	0.49	14.6	0.78	1.44	29.1	157	1110	6.17	55.2	3.71
90	14.0	0.44	14.5	0.71	1.38	26.5	143	999	6.14	49.9	3.70
82	14.3	0.51	10.1	0.86	1.63	24.1	123	882	6.05	29.3	2.48
74	14.2	0.45	10.1	0.79	1.56	21.8	112	796	6.04	26.6	2.48
68	14.0	0.42	10.0	0.72	1.50	20.0	103	723	6.01	24.2	2.46
61	13.9	0.38	10.0	0.65	1.44	17.9	92.2	640	5.98	21.5	2.45
53	13.9	0.37	8.06	0.66	1.44	15.6	77.8	541	5.89	14.3	1.92
48	13.8	0.34	8.03	0.60	1.38	14.1	70.3	485	5.85	12.8	1.91
43	13.7	0.31	8.00	0.53	1.31	12.6	62.7	428	5.82	11.3	1.89
38	14.1	0.31	6.77	0.52	1.06	11.2	54.6	385	5.87	7.88	1.55
34	14.0	0.29	6.75	0.46	1.00	10.0	48.6	340	5.83	6.91	1.53
30	13.8	0.27	6.73	0.39	0.94	8.85	42.0	291	5.73	5.82	1.49
26	13.9	0.26	5.03	0.42	0.94	7.69	35.3	245	5.65	3.54	1.08
22	13.7	0.23	5.00	0.34	0.88	6.49	29.0	199	5.54	2.80	1.04
W 12 x 336	16.8	1.78	13.4	3.00	3.69	98.8	483	4060	6.41	177	3.47
230	15.1	1.29	12.9	2.07	2.75	67.7	321	2420	5.97	115	3.31
120	13.1	0.71	12.3	1.11	1.81	35.3	163	1070	5.51	56.0	3.13
106	12.9	0.61	12.2	0.99	1.69	31.2	145	933	5.47	49.3	3.11
96	12.7	0.55	12.2	0.90	1.63	28.2	131	833	5.44	44.4	3.09
87	12.5	0.52	12.1	0.81	1.50	25.6	118	740	5.38	39.7	3.07
79	12.4	0.47	12.1	0.74	1.44	23.2	107	662	5.34	35.8	3.05
72	12.3	0.43	12.0	0.67	1.38	21.1	97.4	597	5.31	32.4	3.04

TABLE 3-16 DIMENSIONS AND PROPERTIES OF STEEL SECTIONS* (Continued)

"W" & "M" SHAPES

Designation	D, in	T_w, in	B_f, in	T_f, in	K, in	A, in^2	S_x, in^3	I_x, in^4	r_x, in	S_y, in^3	r_y, in
65	12.1	0.39	12.0	0.61	1.31	19.1	87.9	533	5.28	29.1	3.02
58	12.2	0.36	10.0	0.64	1.38	17.0	78.0	475	5.28	21.4	2.51
53	12.1	0.35	10.0	0.58	1.25	15.6	70.6	425	5.23	19.2	2.48
50	12.2	0.37	8.08	0.64	1.38	14.7	64.7	394	5.18	13.9	1.96
45	12.1	0.34	8.05	0.58	1.25	13.2	58.1	350	5.15	12.4	1.94
40	11.9	0.30	8.01	0.52	1.25	11.8	51.9	310	5.13	11.0	1.93
35	12.5	0.30	6.56	0.52	1.00	10.3	45.6	285	5.25	7.47	1.54
30	12.3	0.26	6.52	0.44	0.94	8.79	38.6	238	5.21	6.24	1.52
26	12.2	0.23	6.49	0.38	0.88	7.65	33.4	204	5.17	5.34	1.51
22	12.3	0.26	4.03	0.43	0.88	6.48	25.4	156	4.91	2.31	0.85
19	12.2	0.24	4.01	0.35	0.81	5.57	21.3	130	4.82	1.88	0.82
16	12.0	0.22	3.99	0.27	0.75	4.71	17.1	103	4.67	1.41	0.77
14	11.9	0.20	3.97	0.23	0.69	4.16	14.9	88.6	4.62	1.19	0.75
W 10 x 112	11.4	0.76	10.4	1.25	1.88	32.9	126	716	4.66	45.3	2.68
100	11.1	0.68	10.3	1.12	1.75	29.4	112	623	4.60	40.0	2.65
88	10.8	0.61	10.3	0.99	1.63	25.9	98.5	534	4.54	34.8	2.63
77	10.6	0.53	10.2	0.87	1.50	22.6	85.9	455	4.49	30.1	2.60
68	10.4	0.47	10.1	0.77	1.38	20.0	75.7	394	4.44	26.4	2.59
60	10.2	0.42	10.1	0.68	1.31	17.6	66.7	341	4.39	23.0	2.57

54	10.1	0.37	10.0	0.62	1.25	15.8	60.0	303	4.37	20.6	2.56
49	9.98	0.34	10.0	0.56	1.19	14.4	54.6	272	4.35	18.7	2.54
45	10.1	0.35	8.02	0.62	1.25	13.3	49.1	248	4.32	13.3	2.01
39	9.92	0.32	7.99	0.53	1.13	11.5	42.1	209	4.27	11.3	1.98
33	9.73	0.29	7.96	0.44	1.06	9.71	35.0	170	4.19	9.20	1.94
30	10.5	0.30	5.81	0.51	0.94	8.84	32.4	170	4.38	5.75	1.37
26	10.3	0.26	5.77	0.44	0.88	7.61	27.9	144	4.35	4.89	1.36
22	10.2	0.24	5.75	0.36	0.75	6.49	23.2	118	4.27	3.97	1.33
19	10.2	0.25	4.02	0.40	0.81	5.62	18.8	96.3	4.14	2.14	0.87
17	10.1	0.24	4.01	0.33	0.75	4.99	16.2	81.9	4.05	1.78	0.84
15	9.99	0.23	4.00	0.27	0.69	4.41	13.8	68.9	3.95	1.45	0.81
12	9.87	0.19	3.96	0.21	0.63	3.54	10.9	53.8	3.90	1.10	0.79
W 8 x											
67	9.00	0.57	8.28	0.94	1.44	19.7	60.4	272	3.72	21.4	2.12
58	8.75	0.51	8.22	0.81	1.31	17.1	52.0	228	3.65	18.3	2.10
48	8.50	0.40	8.11	0.67	1.19	14.1	43.3	184	3.61	15.0	2.08
40	8.25	0.36	8.07	0.56	1.06	11.7	35.5	146	3.53	12.2	2.04
35	8.12	0.31	8.02	0.50	1.00	10.3	31.2	127	3.51	10.6	2.03
31	8.00	0.29	8.00	0.44	0.94	9.13	27.5	110	3.47	9.27	2.02
28	8.06	0.29	6.54	0.47	0.94	8.25	24.3	98.0	3.45	6.63	1.62
24	7.93	0.25	6.50	0.40	0.88	7.08	20.9	82.8	3.42	5.63	1.61
21	8.28	0.25	5.27	0.40	0.81	6.16	18.2	75.3	3.49	3.71	1.26
18	8.14	0.23	5.25	0.33	0.75	5.26	15.2	61.9	3.43	3.04	1.23
15	8.11	0.25	4.02	0.32	0.75	4.44	11.8	48.0	3.29	1.70	0.88

TABLE 3-16 DIMENSIONS AND PROPERTIES OF STEEL SECTIONS* *(Continued)*

"W" & "M" SHAPES

Designation		D, in	T_w, in	B_h, in	T_h, in	K, in	A, in²	S_x, in³	I_x, in⁴	r_x, in	S_y, in³	r_y, in
	13	7.99	0.23	4.00	0.26	0.69	3.84	9.91	39.6	3.21	1.37	0.84
	10	7.89	0.17	3.94	0.21	0.62	2.96	7.81	30.8	3.22	1.06	0.84
W 6 x	25	6.38	0.32	6.08	0.46	0.81	7.34	16.7	53.4	2.70	5.61	1.52
	20	6.20	0.26	6.02	0.37	0.75	5.87	13.4	41.4	2.66	4.41	1.50
	16	6.28	0.26	4.03	0.41	0.75	4.74	10.2	32.1	2.60	2.20	0.97
	15	5.99	0.23	5.99	0.26	0.63	4.43	9.72	29.1	2.56	3.11	1.46
	12	6.03	0.23	4.00	0.28	0.63	3.55	7.31	22.1	2.49	1.50	0.92
	9	5.90	0.17	3.94	0.22	0.56	2.68	5.56	16.4	2.47	1.11	0.91
W 5 x	19	5.15	0.27	5.03	0.43	0.81	5.54	10.2	26.2	2.17	3.63	1.28
	16	5.01	0.24	5.00	0.36	0.75	4.68	8.51	21.3	2.13	3.00	1.27
W 4 x	13	4.16	0.28	4.06	0.35	0.69	3.83	5.46	11.3	1.72	1.90	1.00
M 14 x	18	14.0	0.22	4.00	0.27	0.63	5.10	21.1	148	5.38	1.32	0.72
12 x	11.8	12.0	0.18	3.07	0.23	0.56	3.47	12.0	71.9	4.55	0.64	0.53
10 x	9	10.0	0.16	2.69	0.21	0.56	2.65	7.76	38.8	3.83	0.45	0.48
8 x	6.5	8.00	0.14	2.28	0.19	0.50	1.92	4.62	18.5	3.10	0.30	0.42
6 x	20	6.00	0.25	5.94	0.38	0.88	5.89	13.0	39.0	2.57	3.90	1.40

CHANNELS

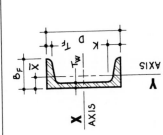

Designation	D, in	T_w, in	B_t, in	T_b, in	K, in	A, in²	S_x, in³	I_x, in⁴	r_x, in	S_y, in³	r_y, in	$\bar{x}$, in
6 x 4.4	6.00	0.17	1.84	0.44		1.29	2.40	7.20	2.36	0.18		0.36
5 x 18.9	5.00	0.42	5.00	0.88		5.55	9.63	24.1	2.08	3.14		1.19
4 x 13	4.00	0.37	3.94	0.81		3.81	5.24	10.5	1.66	1.71		0.94
MC 18 x 58	18.0	0.70	4.20	0.63	1.38	17.1	75.1	676	6.29	5.32	1.02	0.86
51.9	18.0	0.60	4.10	0.63	1.38	15.3	69.7	627	6.41	5.07	1.04	0.86
45.8	18.0	0.50	4.00	0.63	1.38	13.5	64.3	578	6.56	4.82	1.06	0.87
42.7	18.0	0.45	3.95	0.63	1.38	12.6	61.6	554	6.64	4.69	1.07	0.88

TABLE 3-16 DIMENSIONS AND PROPERTIES OF STEEL SECTIONS* *(Continued)*

CHANNELS

Designation	D, in	T_w, in	B_h, in	T_b, in	K, in	A, in²	S_x, in³	I_x, in⁴	r_x, in	S_y, in³	r_y, in	$\bar{x}$, in
C 15 × 50	15.0	0.72	3.72	0.65	1.44	14.7	53.8	404	5.24	3.78	0.87	0.80
40	15.0	0.52	3.52	0.65	1.44	11.8	46.5	349	5.44	3.37	0.89	0.78
33.9	15.0	0.40	3.40	0.65	1.44	9.96	42.0	315	5.62	3.11	0.90	0.79
MC 13 × 50	13.0	0.79	4.41	0.61	1.38	14.7	48.4	314	4.62	4.79	1.06	0.97
40	13.0	0.56	4.19	0.61	1.38	11.8	42.0	273	4.82	4.26	1.08	0.96
35	13.0	0.45	4.07	0.61	1.38	10.3	38.8	252	4.95	3.99	1.10	0.98
31.8	13.0	0.38	4.00	0.61	1.38	9.35	36.8	239	5.06	3.81	1.11	1.00
MC 12 × 50	12.0	0.84	4.14	0.70	1.31	14.7	44.9	269	4.28	5.65	1.09	1.05
45	12.0	0.71	4.01	0.70	1.31	13.2	42.0	252	4.36	5.33	1.09	1.04
40	12.0	0.59	3.89	0.70	1.31	11.8	39.0	234	4.46	5.00	1.10	1.04
35	12.0	0.47	3.77	0.70	1.31	10.3	36.1	216	4.59	4.67	1.11	1.05
C 12 × 30	12.0	0.51	3.17	0.50	1.13	8.82	27.0	162	4.29	2.06	0.76	0.67
25	12.0	0.39	3.05	0.50	1.13	7.35	24.1	144	4.43	1.88	0.78	0.67
20.7	12.0	0.28	2.94	0.50	1.13	6.09	21.5	129	4.61	1.73	0.80	0.70
MC 10 × 41.1	10.0	0.80	4.32	0.58	1.25	12.1	31.5	158	3.61	4.88	1.14	1.09
33.6	10.0	0.58	4.10	0.58	1.25	9.87	27.8	139	3.75	4.38	1.16	1.08

C 10 x 30	10.0	0.67	3.03	0.44	1.00	8.82	20.7	103	3.42	1.65	0.67	0.65
25	10.0	0.53	2.89	0.44	1.00	7.35	18.2	91.2	3.52	1.48	0.68	0.62
20	10.0	0.38	2.74	0.44	1.00	5.88	15.8	78.9	3.66	1.32	0.69	0.61
15.3	10.0	0.24	2.60	0.44	1.00	4.49	13.5	67.4	3.87	1.16	0.71	0.63
MC 9 x 25.4	9.00	0.45	3.50	0.55	1.19	7.47	19.6	88.0	3.43	3.02	1.01	0.97
23.9	9.00	0.40	3.45	0.55	1.19	7.02	18.9	85.0	3.48	2.93	1.01	0.98
C 9 x 20	9.00	0.45	2.65	0.41	0.94	5.88	13.5	60.9	3.22	1.17	0.64	0.58
15	9.00	0.29	2.49	0.41	0.94	4.41	11.3	51.0	3.40	1.01	0.66	0.59
13.4	9.00	0.23	2.43	0.41	0.94	3.94	10.6	47.9	3.48	0.96	0.67	0.60
MC 8 x 20	8.00	0.40	3.03	0.50	1.13	5.88	13.6	54.5	3.05	2.05	0.87	0.84
18.7	8.00	0.35	2.98	0.50	1.13	5.50	13.1	52.5	3.09	1.97	0.87	0.85
C 8 x 13.75	8.00	0.30	2.34	0.39	0.94	4.04	9.03	36.1	2.99	0.85	0.62	0.55
11.5	8.00	0.22	2.26	0.39	0.94	3.38	8.14	32.6	3.11	0.78	0.63	0.52
MC 7 x 22.7	7.00	0.50	3.60	0.50	1.13	6.67	13.6	47.5	2.67	2.85	1.05	1.04
19.1	7.00	0.35	3.45	0.50	1.13	5.61	12.3	43.2	2.77	2.57	1.04	1.08
17.6	7.00	0.38	3.00	0.48	1.06	5.17	10.8	37.6	2.70	1.89	0.88	0.87
C 7 x 14.75	7.00	0.42	2.30	0.37	0.88	4.33	7.78	27.2	2.51	0.78	0.56	0.53
12.25	7.00	0.31	2.19	0.37	0.88	3.60	6.93	24.2	2.60	0.70	0.57	0.53
9.8	7.00	0.21	2.09	0.37	0.88	2.87	6.08	21.3	2.72	0.63	0.58	0.54

TABLE 3-16 DIMENSIONS AND PROPERTIES OF STEEL SECTIONS* *(Continued)*

CHANNELS

Designation		D, in	T_w, in	B_f, in	T_f, in	K, in	A, in²	S_x in³	I_x, in⁴	r_x, in	S_y, in³	r_y, in	$\bar{x}$, in
MC 6 x	18	6.00	0.38	3.50	0.48	1.06	5.29	9.91	29.7	2.37	2.48	1.06	1.12
	15.3	6.00	0.34	3.50	0.39	0.88	4.50	8.47	25.4	2.38	2.03	1.05	1.05
C 6 x	13	6.00	0.44	2.16	0.34	0.81	3.83	5.80	17.4	2.13	0.64	0.53	0.51
	10.5	6.00	0.31	2.03	0.34	0.81	3.09	5.06	15.2	2.22	0.56	0.53	0.50
	8.2	6.00	0.20	1.92	0.34	0.81	2.40	4.38	13.1	2.34	0.49	0.54	0.51
C 5 x	9	5.00	0.33	1.89	0.32	0.75	2.64	3.56	8.90	1.83	0.45	0.49	0.48
	6.7	5.00	0.19	1.75	0.32	0.75	1.97	3.00	7.49	1.95	0.38	0.49	0.48
C 4 x	7.25	4.00	0.32	1.72	0.30	0.69	2.13	2.29	4.59	1.47	0.34	0.45	0.46
	5.4	4.00	0.18	1.58	0.30	0.69	1.59	1.93	3.85	1.56	0.28	0.45	0.46
C 3 x	6	3.00	0.36	1.60	0.27	0.69	1.76	1.38	2.07	1.08	0.27	0.42	0.46
	5	3.00	0.26	1.50	0.27	0.69	1.47	1.24	1.85	1.12	0.23	0.41	0.44
	4.1	3.00	0.17	1.41	0.27	0.69	1.21	1.10	1.66	1.17	0.20	0.40	0.44

ANGLES

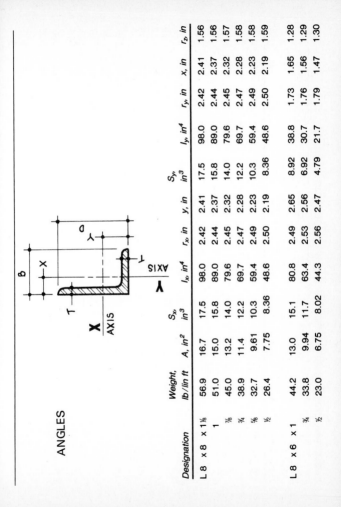

Designation	Weight, lb/lin ft	A, in²	S_x, in³	I_x, in⁴	r_x, in	y, in	S_y, in³	I_y, in⁴	r_y, in	x, in	r_z, in
L 8 × 8 × 1⅛	56.9	16.7	17.5	98.0	2.42	2.41	17.5	98.0	2.42	2.41	1.56
1	51.0	15.0	15.8	89.0	2.44	2.37	15.8	89.0	2.44	2.37	1.56
⅞	45.0	13.2	14.0	79.6	2.45	2.32	14.0	79.6	2.45	2.32	1.57
¾	38.9	11.4	12.2	69.7	2.47	2.28	12.2	69.7	2.47	2.28	1.58
⅝	32.7	9.61	10.3	59.4	2.49	2.23	10.3	59.4	2.49	2.23	1.58
½	26.4	7.75	8.36	48.6	2.50	2.19	8.36	48.6	2.50	2.19	1.59
L 8 × 6 × 1	44.2	13.0	15.1	80.8	2.49	2.65	8.92	38.8	1.73	1.65	1.28
¾	33.8	9.94	11.7	63.4	2.53	2.56	6.92	30.7	1.76	1.56	1.29
½	23.0	6.75	8.02	44.3	2.56	2.47	4.79	21.7	1.79	1.47	1.30

X AXIS

Y AXIS

TABLE 3-16 DIMENSIONS AND PROPERTIES OF STEEL SECTIONS[1] (Continued)

ANGLES

Designation	Weight, lb/lin ft	A, in²	S_x, in³	I_x, in⁴	r_x, in	y, in	S_y, in³	I_y, in⁴	r_y, in	x, in	r_z, in
L 8 x 4 x 1	37.4	11.0	14.1	69.6	2.52	3.05	3.94	11.6	1.03	1.05	0.85
¾	28.7	8.44	10.9	54.9	2.55	2.95	3.07	9.36	1.05	0.95	0.85
½	19.6	5.75	7.49	38.5	2.59	2.86	2.15	6.74	1.08	0.86	0.87
L 6 x 6 x 1	37.4	11.0	8.57	35.5	1.80	1.86	8.57	35.5	1.80	1.86	1.17
⅞	33.1	9.73	7.63	31.9	1.81	1.82	7.63	31.9	1.81	1.82	1.17
¾	28.7	8.44	6.66	28.2	1.83	1.78	6.66	28.2	1.83	1.78	1.17
⅝	24.2	7.11	5.66	24.2	1.84	1.73	5.66	24.2	1.84	1.73	1.18
½	19.6	5.75	4.61	19.9	1.86	1.68	4.61	19.9	1.86	1.68	1.18
⅜	14.9	4.36	3.53	15.4	1.88	1.64	3.53	15.4	1.88	1.64	1.19
L 6 x 4 x ¾	23.6	6.94	6.25	24.5	1.88	2.08	2.97	8.68	1.12	1.08	0.86
⅝	20.0	5.86	5.31	21.1	1.90	2.03	2.54	7.52	1.13	1.03	0.86
½	16.2	4.75	4.33	17.4	1.91	1.99	2.08	6.27	1.15	0.99	0.87
⅜	12.3	3.61	3.32	13.5	1.93	1.94	1.60	4.90	1.17	0.94	0.88
L 5 x 5 x ⅞	27.2	7.98	5.17	17.8	1.49	1.57	5.17	17.8	1.49	1.57	0.97
¾	23.6	6.94	4.53	15.7	1.51	1.52	4.53	15.7	1.51	1.52	0.98
½	16.2	4.75	3.16	11.3	1.54	1.43	3.16	11.3	1.54	1.43	0.98
⅜	12.3	3.61	2.42	8.74	1.56	1.39	2.42	8.74	1.56	1.39	0.99

	9/16	10.3	3.03	2.04	7.42	1.57	1.37	2.04	7.42	1.57	1.37	0.99
L 5 x 3 x	1/2	12.8	3.75	2.91	9.45	1.59	1.75	1.15	2.58	0.83	0.75	0.65
	3/8	9.8	2.86	2.24	7.37	1.61	1.70	0.89	2.04	0.85	0.70	0.65
	5/16	8.2	2.40	1.89	6.26	1.61	1.68	0.75	1.75	0.85	0.68	0.66
	1/4	6.6	1.94	1.53	5.11	1.62	1.66	0.61	1.44	0.86	0.66	0.66
L 4 x 4 x	3/4	18.5	5.44	2.81	7.67	1.19	1.27	2.81	7.67	1.19	1.27	0.78
	5/8	15.7	4.61	2.40	6.66	1.20	1.23	2.40	6.66	1.20	1.23	0.78
	1/2	12.8	3.75	1.97	5.56	1.22	1.18	1.97	5.56	1.22	1.18	0.78
	3/8	9.8	2.86	1.52	4.36	1.23	1.14	1.52	4.36	1.23	1.14	0.79
	5/16	8.2	2.40	1.29	3.71	1.24	1.12	1.29	3.71	1.24	1.12	0.79
	1/4	6.6	1.94	1.05	3.04	1.25	1.09	1.05	3.04	1.25	1.09	0.80
L 4 x 3 x	1/2	11.1	3.25	1.89	5.05	1.25	1.33	1.12	2.42	0.86	0.83	0.64
	3/8	8.5	2.48	1.46	3.96	1.26	1.28	0.87	1.92	0.88	0.78	0.64
	5/16	7.2	2.09	1.23	3.38	1.27	1.26	0.73	1.65	0.89	0.76	0.65
	1/4	5.8	1.69	1.00	2.77	1.28	1.24	0.60	1.36	0.90	0.74	0.65
L 3 x 3 x	1/2	9.4	2.75	1.07	2.22	0.90	0.93	1.07	2.22	0.90	0.93	0.58
	3/8	7.2	2.11	0.83	1.76	0.91	0.89	0.83	1.76	0.91	0.89	0.59
	5/16	6.1	1.78	0.71	1.51	0.92	0.87	0.71	1.51	0.92	0.87	0.59
	1/4	4.9	1.44	0.58	1.24	0.93	0.84	0.58	1.24	0.93	0.84	0.59
	3/16	3.7	1.09	0.44	0.96	0.94	0.82	0.44	0.96	0.94	0.82	0.60

TABLE 3-16 DIMENSIONS AND PROPERTIES OF STEEL SECTIONS* (Continued)

ANGLES

Designation		Weight, lb/lin ft	A, in^2	S_x, in^3	I_x, in^4	r_x, in	y, in	S_y, in^3	I_y, in^4	r_y, in	x, in	r_z, in
L 3 × 2 ×	3/8	5.9	1.73	0.78	1.53	0.94	1.04	0.37	0.54	0.56	0.54	0.43
	5/16	5.0	1.46	0.66	1.32	0.95	1.02	0.32	0.47	0.57	0.52	0.43
	1/4	4.1	1.19	0.54	1.09	0.96	0.99	0.26	0.39	0.57	0.49	0.44
	3/16	3.1	0.90	0.42	0.84	0.97	0.97	0.20	0.31	0.58	0.47	0.44
L 2½ × 2½ ×	3/8	5.9	1.73	0.57	0.98	0.75	0.76	0.57	0.98	0.75	0.76	0.49
	5/16	5.0	1.46	0.48	0.85	0.76	0.74	0.48	0.85	0.76	0.74	0.49
	1/4	4.1	1.19	0.39	0.70	0.77	0.72	0.39	0.70	0.77	0.72	0.49
	3/16	3.1	0.90	0.30	0.55	0.78	0.70	0.30	0.55	0.78	0.70	0.50
L 2½ × 2 ×	3/8	5.3	1.55	0.55	0.91	0.77	0.83	0.36	0.51	0.58	0.58	0.42
	5/16	4.5	1.31	0.47	0.79	0.78	0.81	0.31	0.45	0.58	0.56	0.42
	1/4	3.62	1.06	0.38	0.65	0.78	0.79	0.25	0.37	0.59	0.54	0.42
	3/16	2.75	0.81	0.29	0.51	0.79	0.76	0.20	0.29	0.60	0.51	0.42
L 2 × 2 ×	3/8	4.7	1.36	0.35	0.48	0.59	0.64	0.35	0.48	0.59	0.64	0.39
	5/16	3.92	1.15	0.30	0.42	0.60	0.61	0.30	0.42	0.60	0.61	0.39
	1/4	3.19	0.94	0.25	0.35	0.61	0.59	0.25	0.35	0.61	0.59	0.39
	3/16	2.44	0.72	0.19	0.27	0.62	0.57	0.19	0.27	0.62	0.57	0.39
	1/8	1.65	0.48	0.13	0.19	0.63	0.55	0.13	0.19	0.63	0.55	0.40

Shear

A W 21 x 62 beam of A36 steel is 44 ft long and supports a uniform total load of 1440 lb/lin ft. Is the beam safe in shear?

$$f_v \geq \frac{V}{dt_w}$$

f_v = allowable unit shear for A36 steel (as listed in Table 3-15), 14.5 kips/in^2

V = total shear (half of uniform load), $1440 \times 44/2 = 31.7$ kips

d = depth of beam (from Table 3-16, d of W 21 x 62), 21.0 in

t_w = web thickness of beam (from Table 3-16, t_w of W 21 x 62), 0.40 in

$$f_v = 14.5 \geq \frac{31.7}{20.99 \times 0.4} = 3.77 \text{ kips/in}^2 \quad \text{OK}$$

NOTE: In steel beams, horizontal shear equals vertical shear. In steel sections, only the web area is considered in calculating shear stresses. Practically all rolled W shapes of A36 steel comply with the requirements for compact design.

Bearing Plate Design

A W 21 x 62 beam that has an end reaction of 31.7 kips rests on a bearing plate of A36 steel mounted on a 12-in wide masonry wall (Fig. 3-34). What is the minimum length, width, and thickness of the bearing plate?

STEP 1.   Calculate the minimum plate area.

$$R = AC$$

R = beam end reaction, 31.7 kips

A = minimum area of bearing plate, ? in^2

C = allowable compression stress of wall material, as listed below.

Material	f_c
Masonry or common brick w/cement mortar	0.25 kips/in^2
Sandstone or limestone	0.40 kips/in^2
Concrete, solid reinforced	0.35 f_c
Concrete block, top two rows grouted	0.08 kips/in^2

$$31.7 = A \times 0.25$$
$$A = 127 \text{ in}^2 \text{ minimum area}$$

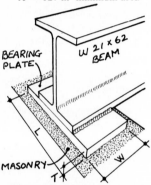

FIGURE 3-34 Steel beam end.

STEP 2. Design the length and width of the bearing plate, taking into consideration the following criteria:

a. The minimum area is 127 in^2

b. The length should be about twice the width.

c. At least 2 in of the material underneath should extend outward beyond all sides of the plate.

One possibility that satisfies the above criteria is $L = 16$ in, $W = 8$ in. (These criteria may also be used to design the width of the supporting wall).

STEP 3. Calculate the bearing plate's thickness.

$$t = \sqrt{\frac{R(L - 2k)^2}{Af_y}}$$

t = minimum thickness of the plate, ? in
R = beam end reaction, 31.7 kips
L = length of plate, 16 in
k = distance from bottom of beam to web toe fillet of W 21 x 62 beam (from Table 3-16), 1.38 in
A = area of the bearing plate, $8 \times 16 = 128$ in^2
f_y = allowable yield stress of bearing plate (from Table 3-15, f_y for A36 steel), 36.0 kips/in^2

$$t = \sqrt{\frac{31.7(16 - 2 \times 1.38)^2}{128 \times 36.0}}$$
$$= 1.10 \text{ in} \quad \text{minimum}$$

Use 1⅛-in thickness.

Bending: Economical Section

A beam of A36 steel supports a uniform load of 1600 lb/lin ft over a span of 30 ft (Fig. 3-35). What is the most economical section for this load? What is the beam's maximum unbraced length?

STEP 1. Find the beam's section modulus.

$$M = f_b S$$

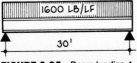

FIGURE 3-35 Beam loading 1.

M = bending moment created by the load; condition is uniform load on a simple span; from Table 3-5:

$$M = WL/8$$

W = weight of total load on beam in kips, 1600 lb $\times$ 30 ft = 48,000 lb = 48 kips

L = length of span in inches, 30 ft $\times$ 12 = 360 in

$$M = \frac{48 \times 360}{8} = 2160 \text{ kip·in}$$

f_b = allowable bending stress for A36 steel (assume compact design; from Table 3-15), 24.0 kips/in^2

S = section modulus of the beam, ? in^3

$$2160 = 24S$$
$$S = 90 \text{ in}^3$$

STEP 2. From Table 3-17 find the economical section for the beam. Under "S_x" locate the smallest number that is more than 90 in^3. It is 94.5. To this number's left is the beam's economical section. It is W 21 x 50.

TABLE 3-17 ECONOMICAL SECTIONS FOR W AND M STEEL SHAPES

Economical section	S_x, in^3	Maximum unbraced length, ft			
		$f_b = 22.0$ kips/in^2	$f_b = 24.0$ kips/in^2	$f_b = 30.0$ kips/in^2	$f_b = 33.0$ kips/in^2
W 36 x 300	1110	35.3	17.6	25.4	14.9
W 36 x 280	1030	33.1	17.5	23.8	14.9
W 36 x 260	953	30.5	17.5	21.9	14.8
W 36 x 245	895	28.6	17.4	20.6	14.8
W 36 x 230	837	26.8	17.4	19.3	14.8
W 33 x 221	757	27.6	16.7	19.8	14.2
W 36 x 210	719	20.9	12.9	15.1	10.9

TABLE 3-17 (Continued)

		Maximum unbraced length, ft			
Economical section	S_x, in^3	$f_b = 22.0$ kips/in^2	$f_b = 24.0$ kips/in^2	$f_b = 30.0$ kips/in^2	$f_b = 33.0$ kips/in^2
W 33 x 201	684	24.9	16.6	17.9	14.1
W 36 x 194	664	19.4	12.8	13.9	10.9
W 36 x 182	623	18.2	12.7	13.1	10.8
W 36 x 170	580	17.0	12.7	12.2	10.8
W 36 x 160	542	15.7	12.7	11.4	10.7
W 36 x 150	504	14.6	12.6	11.3	10.5
W 33 x 141	448	15.4	12.2	11.1	10.3
W 36 x 135	439	13.0	12.3	11.0	8.8
W 33 x 130	406	13.8	12.1	10.8	9.9
W 33 x 118	359	12.6	12.0	10.7	8.6
W 30 x 116	329	13.8	11.1	9.9	9.4
W 30 x 108	299	12.3	11.1	9.8	8.9
W 30 x 99	269	11.4	10.9	9.7	7.9
W 27 x 94	243	12.8	10.5	9.5	8.9
W 24 x 94	222	15.1	9.6	10.9	8.1
W 27 x 84	213	11.0	10.5	9.4	8.0
W 24 x 84	196	13.3	9.5	9.6	8.1
W 24 x 76	176	11.8	9.5	8.6	8.1
W 24 x 68	154	10.2	9.5	8.5	7.4
W 21 x 68	140	12.4	8.7	8.9	7.4
W 24 x 62	131	8.1	7.4	6.4	5.8
W 21 x 62	127	11.2	8.7	8.1	7.4
W 24 x 55	114	7.5	7.0	6.3	5.0
W 18 x 55	98.3	12.1	7.9	8.7	6.7

TABLE 3-17 (Continued)

Economical section	S_x, in^3	Maximum unbraced length, ft			
		$f_b = 22.0$ kips/in^2	$f_b = 24.0$ kips/in^2	$f_b = 30.0$ kips/in^2	$f_b = 33.0$ kips/in^2
W 21 x 50	94.5	7.8	6.9	6.0	5.6
W 18 x 50	88.9	11.0	7.9	7.9	6.7
W 21 x 44	81.6	7.0	6.6	5.9	4.7
W 18 x 40	68.4	8.2	6.3	5.9	5.4
W 16 x 40	64.7	10.2	7.4	7.4	6.3
W 18 x 35	57.6	6.7	6.3	5.6	4.8
W 14 x 34	48.6	10.2	7.1	7.3	6.0
W 16 x 31	47.2	7.1	5.8	5.2	4.9
W 14 x 30	42.0	8.7	7.1	6.5	6.0
W 12 x 30	38.6	10.8	6.9	7.8	5.8
W 16 x 26	38.4	6.0	5.6	5.1	4.0
W 14 x 26	35.3	7.0	5.3	5.1	4.5
W 12 x 26	33.4	9.4	6.9	6.7	5.8
W 14 x 22	29.0	5.6	5.3	4.7	4.1
W 12 x 22	25.4	6.4	4.3	4.6	3.6
W 10 x 22	23.2	9.4	6.1	6.8	5.2
W 12 x 19	21.3	5.3	4.2	3.8	3.6
M 14 x 18	21.1	4.0	3.6	3.4	2.6
W 12 x 16	17.1	4.3	4.1	3.6	2.9
W 12 x 14	14.9	4.2	3.5	3.6	2.5
M 12 x 11.8	12.0	3.0	2.7	2.6	1.9
W 8 x 10	7.81	4.7	4.2	3.7	3.4
M 10 x 9	7.76	2.7	2.6	2.3	1.9
W 6 x 9	5.56	6.7	4.2	4.8	3.5

STEP 3. Find the beam's maximum unbraced length. In Table 3-17, read along the same line as "W 21 x 50" to the number under "f_b = 24.0 kips/in²." It is 6.9 ft.

NOTE: The maximum unbraced length may be extended to the amount listed under "f_b = 22.0 kips/in²" by using f_b = 22.0 kips/in² to recalculate S_x. In this case, S_x = 2160/22 = 98.2 in³. In Table 3-17 the economical section for this amount is W 18 x 55. The maximum unbraced length for this beam is listed on the same line under "f_b = 22.0 kips/in²." It is 12.1 ft.

Bending: Stud Decking Design

A 30-ft span with a total uniform load of 1600 lb/lin ft is covered with a 4-in concrete slab on 2 x 6 metal decking as shown in Fig. 3-36. The

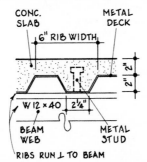

CONC. SLAB

METAL DECK

6" RIB WIDTH

W 12 x 40

2 ¼"

BEAM WEB

METAL STUD

RIBS RUN ⊥ TO BEAM

FIGURE 3-36 Stud decking cross section.

top of the beam is lined with ¾ x 3 in welded steel studs, the beam is compact, and the allowable stress of the concrete is 3.0 kips/in². What size beam should be used, and what should be the spacing of the studs?

STEP 1. Find the maximum bending moment, then use this to obtain what is known as the transformed section modulus. Use this value to find the beam size from Table 3-18.

from Table 3-5 $$M = \frac{WL}{8} = f_b S$$

M = maximum bending moment
W = weight of load on beam, $1600 \times 30 = 48{,}000$ lb
L = length of load on beam, $30 \times 12 = 360$ in
f_b = allowable bending stress for A36 steel and compact design, from Table 3-15, 24,000 lb/in^2
S = section modulus, ? in^3

$$\frac{48{,}000 \times 360}{8} = 24{,}000S$$
$$S = 90 \text{ in}^3$$

Beam size: In Table 3-18 under "Slab thickness, 4 in," find the lowest section modulus value above 90. It is 92.9. Read to the left to find the beam size. It is W 16 x 40.

TABLE 3-18   ECONOMICAL SECTIONS FOR COMPOSITE DESIGN

Economical section	Slab thickness, in			Economical section	Slab thickness, in		
	4	5	6		4	5	6
W 36 x 170	713	748	781	W 14 x 48	99.6	109	120
W 36 x 160	670	702	733	W 18 x 40	99.4	108	116
W 36 x 150	628	659	688	W 16 x 40	92.9	101	110
W 33 x 152	603	633	661	W 14 x 43	89.3	98.1	108
W 33 x 141	559	587	613	W 18 x 35	86.0	93.4	101
W 36 x 135	558	586	612	W 16 x 36	82.8	90.4	98.4
W 33 x 130	512	538	562	W 14 x 38	80.4	88.5	97.0
W 30 x 132	481	506	531	W 14 x 34	72.1	79.3	87.0
W 33 x 118	461	484	506	W 16 x 31	71.9	77.9	84.9
W 30 x 116	421	443	465	W 12 x 35	68.6	76.1	84.1

TABLE 3-18 (Continued)

Economical section	Slab thickness, in			Economical section	Slab thickness, in		
	4	5	6		4	5	6
W 30 x 108	388	409	429	W 14 x 30	63.6	70.1	76.9
W 30 x 99	354	373	392	W 16 x 26	59.7	65.4	71.3
W 27 x 94	316	334	351	W 12 x 30	58.8	65.3	72.1
W 24 x 94	290	307	325	W 14 x 26	55.2	60.9	66.9
W 27 x 84	282	297	313	W 12 x 26	51.4	57.1	63.1
W 24 x 84	259	274	290	W 14 x 22	46.5	51.4	56.5
W 24 x 76	234	248	262	W 12 x 22	42.7	47.7	52.8
W 24 x 68	209	221	234	W 10 x 22	38.8	43.8	49.0
W 21 x 68	189	202	215	W 8 x 24	36.7	42.3	48.2
W 24 x 62	184	196	209	W 12 x 19	36.7	41.0	45.5
W 21 x 62	173	185	197	W 10 x 19	33.4	37.8	42.4
W 24 x 55	164	175	186	W 8 x 21	32.7	37.6	42.7
W 21 x 57	156	167	179	W 12 x 16	30.8	34.5	38.3
W 18 x 60	149	160	173	W 10 x 17	29.5	33.5	37.6
W 18 x 55	137	147	159	W 8 x 18	28.0	32.2	36.6
W 21 x 50	136	146	157	W 10 x 15	26.0	29.5	33.2
W 18 x 50	125	134	145	W 8 x 15	23.5	27.1	30.8
W 21 x 44	120	129	138	W 10 x 12	21.0	23.9	26.8
W 18 x 46	114	123	133	W 8 x 13	20.3	23.4	26.7
W 16 x 45	104	113	123	W 8 x 10	15.9	18.4	20.9

STEP 2. Find the total horizontal shear acting through the studs on the concrete slab and the steel beam flange. The lower value governs.

Steel:

$$v_h = 0.5 A f_y$$

Concrete:

$$v_h = 0.425f_c d(16t + b_f)$$

v_h = total horizontal shear (minimum value governs), ? kips

A = cross-sectional area of beam (from Table 3-16, A for W 16 x 40), 11.8 in^2

f_y = yield stress of steel beam (for A36 steel), from Table 3-15, 36.0 kips/in^2

f_c = compressive stress of concrete, 3.0 kips/in^2

d = minimum depth of concrete slab, 2 in

t = total thickness of concrete slab, 4 in

b_f = flange width of beam under ribbing (from Table 3-16, b_f for W 16 x 40), 7.00 in

Steel:

$$v_h = 0.5 \times 11.8 \times 36.0 = 212 \text{ kips}$$

Concrete:

$$v_h = 0.425 \times 3.0 \times 2(16 \times 4 + 7) = 181 \text{ kips}+$$

The minimum value governs:

$$v_h = 181 \text{ kips}$$

STEP 3. Calculate the number of required studs.

$$v_h = \frac{0.425NLPW}{H}\left(\frac{T}{H} - 1\right)$$

v_h = total horizontal shear from step 2, 181 kips

N = number of studs along total length of beam, ?

L = allowable shear load on each stud, as listed below:

Stud size	f_c = 3.0 kips/in^2	f_c = 4.0 kips/in^2
½-in dia., 2-in shank including head	5.1	5.9

⅝-in dia., 2½-in shank including head	8.0	9.2
¾-in dia., 3-in shank including head	11.5+	13.3
⅞-in dia., 3½-in shank including head	15.6	18.0

P = parallel/perpendicular factor; if ribs run parallel to beam, P = 0.7; if ribs run perpendicular to beam, P = 1.0; in this case, ribs run perpendicular to beam, P = 1.0

W = width of decking rib, 2¼ in = 2.25 in

H = height of decking, 2 in

T = thickness of concrete slab and decking, 4 in

$$181 = \frac{0.425 \times N \times 11.5 \times 1.0 \times 2.25}{2}\left(\frac{4}{2} - 1\right)$$

$$N = 32.9 \qquad \text{Use 33 studs}$$

Spacing:

$$\frac{\text{Length of span}}{\text{Number of studs}} = \frac{30 \times 12}{33} = 10.9 \text{ in} \qquad \text{Use 10¾ in}$$

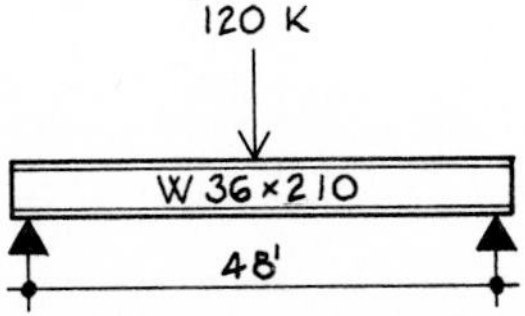

FIGURE 3-37 Beam loading 2.

Deflection

A W 36 x 210 girder is loaded as shown in Fig. 3-37. A plaster ceiling exists on the underside. Is this beam safe in deflection?

From Table 3-5, the formula for maximum deflection of a simple beam, concentrated load at center, is

$$\Delta = \frac{PL^3}{48EI}$$

Δ = maximum deflection for live load, from Table 3-10:

  Plaster or drywall ceilings, $\Delta \leq 1/360$ span
  Other ceilings, $\Delta \leq 1/240$ span

P = point load on the beam, 120 kips = 120,000 lb
L = length of span, 48 × 12 = 576 in
E = modulus of elasticity for steel, 29,000,000 lb/in^2
I = moment of inertia for beam section; from Table 3-16, I_x for W 36 x 210 = 13,200 in^4

$$\text{Actual } \Delta = \frac{120,000 \times 576^3}{48 \times 29,000,000 \times 13,200} = 1.25 \text{ in}$$

$$\text{Allow } \Delta \leq \frac{L}{360} = \frac{576}{360} = 1.25 \leq 1.60 \quad \text{OK}$$

Open Web Steel Joists

Open web joists (H series) at 4-ft centers and 40 ft long support a department store roof that has a dead load of 50 lb/ft^2 and a live load of 30 lb/ft^2. What is the lightest H section that can safely support the load?

STEP 1. Find the total uniform load on each joist.

$$W = S(L + 0.67)(LL + DL)$$

W = weight of total uniform load on each joist, ? lb
S = joist spacing in feet (joists are 4 ft on centers), 4.0 ft
L = length of joist in feet, 40 ft
LL = live load resting on joists, 50 lb/ft^2
DL = dead load resting on joists, 30 lb/ft^2

$$W = 4.0(40 + 0.67)(50 + 30)$$
$$= 13,000 \text{ lb}$$

STEP 2.   Find the required section modulus of the joist.

$$\frac{WL}{8} = f_b S$$

W = weight of total uniform load upon the joist, 13,000 lb
L = length of joist in inches, 40 ft $\times$ 12 = 480 in
f_b = allowable stress for open web steel joists, as listed below:

> J series joists, f_b = 22,000 lb/in^2
> H series joists, f_b = 30,000 lb/in^2

These joists are H series; therefore f_b = 30,000 lb/in^2
S = required section modulus of each joist, ? in^3

$$\frac{13,000 \times 480}{8} = 30,000S$$

$$S = 26.0 \text{ in}^3$$

STEP 3.   Find the economical section in Table 3-19. Under "S_x" in the H series find the smallest section modulus number that is greater than 26.0, then proceed to the left to the column under "Joist Size." This is the lightest open web joist that will carry the load. The smallest section modulus number above 26.0 is 26.1. Its economical section is 26 H 8, which weighs 12.8 lb/lin ft.

TABLE 3-19   ECONOMICAL SECTIONS FOR OPEN WEB STEEL JOISTS

Joist size	S_x, in^3	Weight, lb/ft	Max. end reaction, lb	Depth, in	Limit of span, ft	Check shear, ft or less	Roofs. only, ft or more
			"J" SERIES				
72 DLJ 20	648	83	37,000	72	84–144	128	
68 DLJ 20	610	83	37,100	68	80–136	120	

TABLE 3-19 *(Continued)*

Joist size	S_x, in^3	Weight, lb/ft	Max. end reaction, lb	Depth, in	Limit of span, ft	Check shear, ft or less	Roofs. only, ft or more
64 DLJ 20	572	83	37,200	64	75–128	112	
72 DLJ 19	537	72	30,400	72	84–144	128	
68 DLJ 19	513	72	31,000	68	80–136	120	
64 DLJ 19	487	72	31,500	64	75–128	112	
60 DLJ 19	460	72	32,000	60	70–120	104	
72 DLJ 18	455	62	25,900	72	84–144	128	
68 DLJ 18	429	62	26,000	68	80–136	120	
72 DLJ 17	394	55	22,500	72	84–144	128	
68 DLJ 17	372	55	22,600	68	80–136	120	
64 DLJ 17	349	55	22,700	64	75–128	112	
72 DLJ 16	342	49	19,500	72	84–144	128	
68 DLJ 16	323	49	19,600	68	80–136	120	
64 DLJ 16	302	49	19,800	64	75–128	112	
68 DLJ 15	293	44	17,700	68	80–136	128	
64 DLJ 15	280	44	18,100	64	75–128	112	
60 DLJ 15	262	44	18,400	60	70–120	104	
56 DLJ 15	244	44	18,600	56	66–112	96	
64 DLJ 14	235	39	15,300	64	75–128	112	
56 DLJ 13	190	35	14,400	56	66–112	96	
52 DLJ 12	156	31	12,800	52	61–104	88	
48 LJ 12	123	30	11,000	48	56–96	80	81
44 LJ 12	117	30	11,600	44	52–88	72	74
40 LJ 12	110	30	12,200	40	47–80	64	67

TABLE 3-19 (Continued)

Joist size	S_x, in³	Weight, lb/ft	Max. end reaction, lb	Depth, in	Limit of span, ft	Check shear, ft or less	Roofs. only, ft or more
36 LJ 12	97.6	30	12,600	56	42–72	56	61
40 LJ 11	94.3	28	10,400	40	47–80	64	67
44 LJ 10	90.8	26	8900	44	52–88	72	74
40 LJ 10	86.4	26	9500	40	47–80	64	67
36 LJ 10	79.5	26	10,100	36	42–72	56	61
40 LJ 09	78.1	23	8600	40	47–80	64	67
36 LJ 09	71.0	23	9150	36	42–72	56	61
36 LJ 08	62.6	22	7850	36	42–72	56	61
32 LJ 08	58.1	22	8450	32	38–64	48	54
28 LJ 08	52.6	22	9100	28	33–56	40	47
32 LJ 07	51.1	20	7400	32	38–64	48	54
30 J 11	46.8	18.3	7400	30	30–60	46	
28 J 11	43.5	17.9	7100	28	28–56	44	
30 J 10	40.2	16.6	6800	30	30–60	43	
28 J 10	37.5	15.9	6600	28	28–56	41	
30 J 9	36.0	14.9	6400	30	30–60	41	
28 J 9	33.5	14.3	6200	28	28–56	39	
26 J 9	31.0	14.1	5900	26	26–52	38	
30 J 8	30.3	13.8	5900	30	30–60	37	
24 J 9	28.5	13.3	5600	24	24–48	37	
28 J 8	28.2	13.0	5700	28	28–56	36	
26 J 8	26.1	12.2	5400	26	26–52	35	
22 J 8	22.4	11.9	4800	22	22–44	34	
24 J 7	20.9	11.1	4700	24	24–48	32	

TABLE 3-19 (Continued)

Joist size	S_x, in³	Weight, lb/ft	Max. end reaction, lb	Depth, in	Limit of span, ft	Check shear, ft or less	Roofs. only, ft or more
22 J 7	19.1	10.5	4500	22	22–44	31	
24 J 6	16.7	9.9	4400	24	24–48	27	
22 J 6	15.2	9.6	4200	22	22–44	26	
20 J 6	14.4	9.2	4100	20	20–40	25	
18 J 6	13.3	9.0	3900	18	18–36	25	
20 J 5	12.0	8.1	3800	20	20–40	23	
18 J 5	11.0	7.9	3500	18	18–36	23	
16 J 5	9.82	7.6	3300	16	16–32	21	
14 J 5	8.64	7.3	3100	14	14–28	20	
16 J 4	7.86	6.6	3000	16	16–32	19	
14 J 4	7.23	6.4	2800	14	14–28	18	
12 J 4	6.14	6.0	2500	12	12–24	17	
14 J 3	5.77	5.2	2400	14	14–28	17	
12 J 3	4.91	5.1	2300	12	12–24	15	
10 J 3	4.05	4.8	2200	10	10–20	13	
8 J 3	3.18	4.8	2000	8	8–16	11	
				"H" SERIES			
72 DLH 19	515	70	40,100	72	84–144	128	
68 DLH 19	485	70	40,200	68	80–136	120	
72 DLH 18	442	62	34,200	72	84–144	128	
68 DLH 18	424	62	35,000	68	80–136	120	
64 DLH 18	402	62	35,800	64	75–128	112	

TABLE 3-19 (Continued)

Joist size	S_x, in^3	Weight, lb/ft	Max. end reaction, lb	Depth, in	Limit of span, ft	Check shear, ft or less	Roofs. only, ft or more
72 DLH 17	379	55	29,200	72	84–144	128	
68 DLH 17	367	55	30,200	68	80–136	120	
64 DLH 17	349	55	31,000	64	75–128	112	
72 DLH 16	337	50	22,500	72	84–144	128	
68 DLH 16	323	50	28,600	68	80–136	120	
64 DLH 16	303	50	26,900	64	75–128	112	
72 DLH 15	285	43	22,500	72	84–144	128	
68 DLH 15	272	43	22,600	68	80–136	120	
72 DLH 14	254	39	19,600	72	84–144	128	
68 DLH 14	246	39	20,200	68	80–136	120	
64 DLH 14	235	39	20,800	64	75–128	112	
60 DLH 14	220	39	21,000	60	70–120	104	
68 DLH 13	214	36	17,500	68	80–136	120	
64 DLH 13	207	36	18,200	64	75–128	112	
60 DLH 13	199	36	18,900	60	70–120	104	
56 DLH 13	190	36	19,600	56	66–112	96	
52 DLH 13	176	36	19,800	52	61–104	88	
64 DLH 12	170	31	15,000	64	75–128	112	
60 DLH 12	164	31	15,500	60	70–120	104	
56 DLH 12	156	31	16,200	56	66–112	96	
52 DLH 12	145	31	16,400	52	61–104	88	
56 DLH 11	137	29	14,000	56	66–112	96	
52 DLH 11	136	29	14,600	52	61–104	88	

TABLE 3-19 (Continued)

Joist size	S_x, in³	Weight, lb/ft	Max. end reaction, lb	Depth, in	Limit of span, ft	Check shear, ft or less	Roofs. only, ft or more
52 DLH 10	118	27	13,400	52	61–104	88	
48 LH 11	90.3	27	10,800	48	56–96	80	
44 LH 11	89.8	27	12,000	44	52–88	72	74
40 LH 11	87.6	27	13,100	40	47–80	64	67
48 LH 10	83.4	25	10,000	48	56–96	80	81
44 LH 10	82.9	25	11,000	44	52–88	72	74
40 LH 10	79.7	25	12,000	40	47–80	64	67
44 LH 09	75.1	22	10,000	44	52–88	72	74
36 LH 08	56.4	20	9250	36	42–72	56	61
40 LH 08	55.7	20	8300	40	47–80	64	67
36 LH 07	51.6	20	8400	36	42–72	56	61
32 LH 07	47.7	20	9400	52	38–64	48	54
28 H 11	43.3	18.3	8400	28	28–56	46	52
32 LH 06	42.6	18.0	8400	32	38–64	48	54
26 H 11	40.1	17.9	8300	26	26–52	48	44
28 H 10	37.5	16.8	7700	28	28–56	48	47
30 H 9	35.8	15.4	7500	30	30–60	47	51
28 H 9	33.3	15.2	7200	28	28–56	46	47
26 H 9	30.8	14.8	7200	26	26–52	42	44
30 H 8	30.3	14.2	6800	30	30–60	44	51
24 H 9	28.4	14.0	7000	24	24–48	40	41
28 H 8	28.2	13.5	6700	28	28–56	46	43

TABLE 3-19 (Continued)

Joist size	S_x, in^3	Weight, lb/ft	Max. end reaction, lb	Depth, in	Limit of span, ft	Check shear, ft or less	Roofs. only, ft or more
26 H 8	26.1	12.8	6700	26	26–52	39	44
24 H 8	23.9	12.7	6000	24	24–48	39	41
22 H 8	21.8	12.0	5800	22	22–44	36	38
24 H 7	19.2	11.5	5800	24	24–48	33	41
22 H 7	17.5	10.7	5600	22	22–44	31	37
20 H 7	16.6	10.7	5400	20	20–40	30	34
18 H 7	15.5	10.4	5200	18	18–36	29	31
24 H 6	15.4	10.3	5600	24	24–48	27	41
22 H 6	14.1	9.7	5400	22	22–44	26	37
20 H 6	13.5	9.6	5100	20	20–40	26	34
18 H 6	12.8	9.2	4800	18	18–36	26	31
20 H 5	12.2	8.4	4800	20	20–40	25	34
18 H 5	10.8	8.0	4500	18	18–36	24	31
16 H 5	9.63	7.8	4300	16	16–32	22	27
14 H 5	8.63	7.4	3800	14	14–28	22	24
12 H 5	7.40	7.1	3600	12	12–24	20	21
16 H 4	7.37	6.6	3800	16	16–32	19	27
14 H 4	7.07	6.5	3500	14	14–28	20	24
12 H 4	6.00	6.2	3200	12	12–24	18	21
14 H 3	5.5	5.5	3200	14	14–28	17	24
12 H 3	4.67	5.2	2800	12	12–24	16	21
10 H 3	3.87	5.0	2500	10	10–20	15	17
8 H 3	3.03	5.0	2400	8	8–16	12	14

NOTE: Open web steel joists are designated by the joist depth, series, and top chord section number, as indicated below. (See Fig. 3-38.)

FIGURE 3-38   Open web steel joist label.

Joist depth   Given in multiples of 2 in from 8 to 72 in.

Series   This information is listed below. H series joists are made of higher strength steel and hence have greater spanning capabilities.

Top chord section number   Each number refers to a standard section used by that manufacturer. For example, CECO's top chord section number 5 is two angles 1.50 × 1.50 × 0.138 in in size. Another manufacturer's section number 5 may be a slightly different angle having slightly different properties, and thus its joist tables may be slightly different.

Series	Description	f_b, kips/in²	Depth, in	Span, ft
J	Standard	22.0	8 to 30	8 to 60
H	Standard	30.0	8 to 30	8 to 60
LJ	Longspan	22.0	18 to 48	25 to 96
LH	Longspan	30.0	18 to 48	25 to 96
DLJ	Deep longspan	22.0	52 to 72	88 to 144
DLH	Deep longspan	30.0	52 to 72	88 to 144

Columns: Axial Loads

A W 10 x 33 column of A36 steel has an unbraced length of 12 ft. The column has pinned ends ($K = 1.0$). What is its maximum load?

STEP 1.   Compute the column's slenderness ratio.

$$R = \frac{KL}{r}$$

R = slenderness ratio, **?**

K = column end condition factor, 1.0

L = unbraced length of column, 12 ft = 144 in

r = radius of gyration of column cross section (from Table 3-16, r_x = 4.19 in, r_y = 1.94 in, use the smaller value), 1.94 in

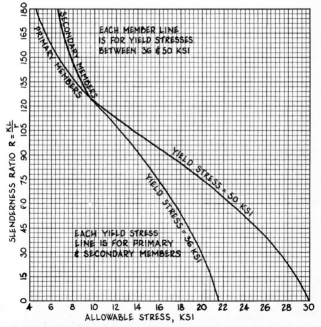

FIGURE 3-38a Allowable stresses for compression members.

$$R = \frac{1.0 \times 144}{1.94} = 74.2 \qquad \text{Use 74}$$

STEP 2.   Find the allowable stress for a slenderness ratio of 74 in Fig. 3-38a. $f_a = 15.9$ kips/in^2.

STEP 3.   Compute the maximum allowable load.

$$P = Af_a$$

P = maximum allowable load on the column, ? kips

A = cross-sectional area of column (from Table 3-16, A for W 10 x 33), 9.71 in^2

f_a = allowable unit stress from step 2, 16.0 kips/in^2

$$P = 9.71 \times 16.0$$
$$= 155 \text{ kips}$$

End Condition Analysis

A W 14 x 61 column is braced as shown in Fig. 3-39. What are the slenderness ratios for each axis, and which one governs design?

STEP 1.   Find r_x and r_y from Table 3-16.

$$r_x = 5.98 \text{ in}$$
$$r_y = 2.45 \text{ in}$$

STEP 2.   Find the column K values from Table 3-6.

K about X axis:

Top: rotation, fixed
    translation, free   } $K = 2.0$
Bottom rotation, free

K about Y axis:

Top: rotation, free
    translation, fixed   } $K = 0.80$
Bottom rotation, fixed

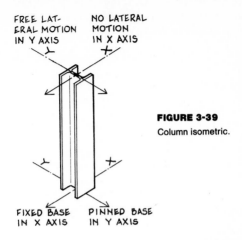

FREE LAT-
ERAL MOTION
IN Y AXIS

NO LATERAL
MOTION
IN X AXIS

FIGURE 3-39
Column isometric.

FIXED BASE
IN X AXIS

PINNED BASE
IN Y AXIS

STEP 3. Compute the slenderness ratio in both axes. The smaller value governs.

$$R = \frac{KL}{r} \qquad R_x = \frac{2.0 \times 138}{5.98} \qquad R_y = \frac{0.80 \times 138}{2.45}$$
$$= 46.1 \qquad\qquad = 45.1 +$$

The slenderness ratio about the Y axis governs design.

Column Design

A 12 ft tall column supports an axial load of 80 kips. What is the lightest structural steel shape of A36 steel that will safely support the load? $K = 1.0$.

STRATEGY For this problem you will need the AISC Manual, 8th ed. Scan the column load tables from AISC page 3-16 to 3-40. Note all cross sections that will support 80 kips or more at a $L_u = 12$ ft and check their weights per linear foot. The best ones are

AISC page	Structural shape	Weight, lb/lin ft
3-29	W 6 x 25	25.0
3-29	W 6 x 20	20.0
3-34	Standard steel pipe, 6-in dia., 0.280-in wall	19.0
3-35	Extra strong pipe, 5-in dia., 0.375-in wall	20.8
3-39	Square structural tubing, 6 x 6 in, ³⁄₁₆-in wall	14.5
3-39	Square structural tubing, 5 x 5 in, ¼-in wall	15.6

The winner is 6 x 6 in square structural tubing with walls ³⁄₁₆-in thick.

Built-up Steel Sections

The built-up section shown in Fig. 3-40 is fabricated from three C 6 x 13 channels and a ½ x 8 in steel plate. How far is the center axis of this section from the outer face of the steel plate?

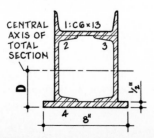

FIGURE 3-40 Built-up steel cross section.

STRATEGY   Take moments about the bottom of the built-up section. Sum of the moments of the parts equals the moment of the whole. Moment equals area of each section times moment arm. (See Fig. 3-41.)

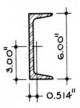

FIGURE 3-41   Channel cross section.

$$M_1 + M_2 + M_3 + M_4 = M_t$$
$$\text{Each } M = Ad$$
$$A_1d_1 + A_2d_2 + A_3d_3 + A_4d_4 = A_tD$$

$A_1 = A_2 = A_3 =$ cross-sectional area of channels 1, 2, and 3 (from Table 3-16), $A = 3.83 \text{ in}^2$

$d_1 =$ moment arm distance of channel 1, $0.514 + 6.00 + 0.50 = 7.01$ in (relevant dimensions for the channels found in Table 3-16)

$d^2 = d_3 =$ moment arm distances of channels 2 and 3, $3.00 + 0.50 = 3.50$ in

$A_4 =$ cross-sectional area of plate, $0.50 \times 8 = 4.00 \text{ in}^2$

$d_4 =$ moment arm of plate, $0.50/2 = 0.25$ in

$A_t =$ total area of section, equals $A_1 + A_2 + A_3 + A_4 = 3.83 \times 3 + 4.00 = 15.5 \text{ in}^2$

$D =$ distance of center axis to outer face of plate, **?** in

$$3.83 \times 7.01 + 3.83 \times 3.50 \times 2 + 4.0 \times 0.25 = 15.5D$$
$$D = 3.53 \text{ in}$$

Combined Compression and Bending

A 16-ft tall column of A572-50 steel supports an axial load of 320 kips and has a moment of 84 ft·kips about its Y axis. Select the economical section. $K = 1.0$.

STEP 1.   Try a W-shape size in the formula below. The most economical section will have a total value just under 1.0.

$$\frac{P}{Af_a} + \frac{M}{Sf_b} \le 1.0 \qquad \text{Try a W 16 x 67}$$

P = point load of the column, 320 kips

A = cross sectional area of the column (from Table 3-16, A for a W 16 × 67), 19.7 in^2

f_a = allowable unit stress, found by calculating the slenderness ratio R:

$$R = KL/r$$

K = column end condition, 1.0

L = unbraced length of column in inches, 16 x 12 = 192 in

r = radius of gyration of column cross section; from Table 3-16, r_y for W 16 × 57 = 246

$$R = 1.0 \times 192/2.46 = 78.0$$

from Fig. 3-38a, when $R = 78.0$ for A572-50 steel (from Table 3-15, f_y = 50 kips/in^2), f_a = 19.4 kips/in^2

M = maximum bending moment in column in in·kips, 84 ft·kips × 12 = 1010 in·kips

S = section modulus of W 16 x 67 about the Y axis (from Table 3-16), S_y = 23.2 in^3

f_b = allowable bending stress for A572-50 steel (from Table 3-15), 30.0 kips/in^2

$$\frac{320}{19.7 \times 19.4} + \frac{1010}{23.2 \times 30} = 2.29$$

$$2.29 \gg 1.0 \qquad \text{NG} \qquad \text{Much too high}$$

STEP 2.   Try larger W shapes. Although 14 W shapes are narrower than 16 W shapes, many are designed especially for columnar loading. Try these first.

Try a W 14 x 90. Above format yields 1.18   NG
Try a W 14 x 120. Above format yields 0.87   OK, but low
Try a W 14 x 109. Above format yields 0.97   OK
Try a W 18 x 106. Above format yields 1.35   NG
Try a W 24 x 104. Above format yields 1.31   NG

Use W 14 x 109.

Column Pedestal Design

A W 10 x 45 column supporting a total load of 270 kips bears on a steel base plate mounted on a concrete pedestal. The steel is A36 and the concrete stress is 3 kips/in². What is the minimum length, width, and thickness of the base plate?

STEP 1.   Compute the required area of the bearing plate.

$$P = 0.25Af_c$$

P = point load on the column, 270 kips
A = minimum surface area of the base plate, ? in²
f_c = allowable unit stress of the concrete, 3 kips/in²

$$270 = 0.25A \times 3$$
$$A = 360 \text{ in}^2$$

Use 18 x 20 in base.

STEP 2.   Find M and N in Fig. 3-42. The larger value governs.

$$M = \frac{L - 0.95d}{2}$$
$$N = \frac{W - 0.80b}{2}$$

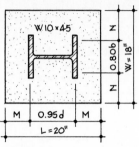

FIGURE 3-42 Column pedestal.

L = length of bearing plate, 20 in
W = width of bearing plate, 18 in
d = depth of W 10 x 45 (from Table 3-16), 10.1 in
b = flange width of W 10 x 45 (from Table 3-16), 8.02 in

$$M = \frac{20 - 0.95 \times 10.1}{2} = 5.20 \text{ in}$$

$$N = \frac{18 - 0.80 \times 8.02}{2} = 5.79 \text{ in} + \qquad \text{Use this number below}$$

STEP 3. Calculate the thickness of the bearing plate.

$$t = \frac{f_c N^2}{f_y}$$

t = thickness of the bearing plate, ? in
f_c = allowable unit stress of the concrete, 3 kips/in^2
N = from step 2, 5.79 kips/in^2
f_y = yield stress for A36 steel, 36 kips/in^2

$$t = \frac{3 \times 5.79^2}{36} = 1.67 \text{ in} \qquad \text{Use 1¾ in}$$

Base plate size: 1¾ x 18 x 20 in

Bolt Connection

Example 1   What is the allowable load for the joint shown in Fig. 3-43 if 1-in diameter ASTM A490 steel bolts are used? The plates are A36 steel.

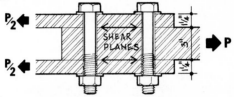

P/2 ←

P/2 ←

SHEAR PLANES

1¼"

3"

1¼"

→ P

FIGURE 3-43   Steel joint showing shear.

STRATEGY   Solve for the amount of shear passing through the bolt, then the bearing value of bolt against plate. The lower value governs.

STEP 1.   Establish the bolt type for the connection.

Type of steel, A490
Type of connection: There are three kinds, as shown in Fig. 3-44.

a.   *Friction type* (F):   The clamping action of the tightened high-tensile-stress bolt creates friction between the connected parts, which resists the applied load.

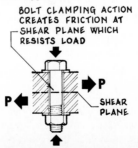

BOLT CLAMPING ACTION
CREATES FRICTION AT
SHEAR PLANE WHICH
RESISTS LOAD

→ P

P ←

SHEAR PLANE

b. *Bearing type, threads included* (N): The connected pieces press against the sides of the tightened bolt whose threads exist within the connection's shear plane.

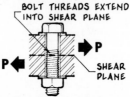

c. *Bearing type, threads excluded* (X): The connected pieces press against the side of the tightened bolt whose threads do not exist in the connection's shear plane.

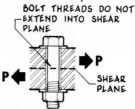

FIGURE 3-44 Bolt in steel joint.

Inspection of the joint reveals two types of connection: the upper shear plane is *bearing, threads excluded* (X) and the lower shear plane is *bearing, threads included* (N). Thus the upper connection is A490 X and the lower connection is A490 N.

STEP 2. Find the amount of shear passing through the bolt. In Fig. 3-44a look up the connection in one shear plane at a time. The bolt diameter is 1 in. For the upper connection, find "A490 X" in the column on the left, then follow the solid line at its right downward into the graph until the line meets the vertical "1-in" diameter line. From this intersection, read upward to the right to obtain the connection's allowable load in kips. It is 31.5 kips. For the lower connection, find "A490 N" in the column on the left, then in like manner find the connection's allowable load. It is 22.0 kips.

The total allowable shear that can pass through each bolt is 31.5 + 22.0 = 53.5 kips. The total allowable load for both bolts is 53.5 × 2 = 107 kips.

STEP 3.　Compute the bearing value of bolt to plate.

$$P = Nf_cDT$$

P = allowable point load, ? kips
N = number of bolts, 2
f_c = allowable unit compression stress for A36 steel (from Table 3-15), 22.0 kips/in^2
D = diameter of bolts, 1 in
T = thickness of plate; use the lower amount below:
　　T for the inner plate, 3 in
　　T for the two outer plates, 1.25 × 2 = 2.50 in +

$$P = 2 \times 22.0 \times 1 \times 2.50$$
$$= 110 \text{ kips bearing load}$$

STEP 4.　Compare the amount of shear passing through the bolt with the bearing value of bolt against plate. The lower value governs.

The 107 kips shear force is less than the 110 kips bearing value. Thus the allowable load for the joint is 107 kips.

Example 2　In the problem above, what is the allowable load for the joint if the connection has a friction type fitting? The holes through the plates are ⅟₁₆ inch larger than the bolt diameters, which means they are standard round holes.

STRATEGY　Calculate the connection's shear load from Fig. 3-44a, then compare the amount with the connection's bearing load, which is the same as in the previous problem.

FOR STANDARD-SIZE HOLES (HOLE DIA. IS $\frac{1}{16}$" LARGER THAN BOLT DIA.)

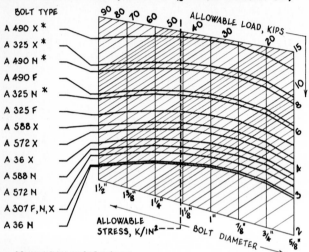

CONNECTION DESCRIPTIONS...

F = FRICTION TYPE

N = BEARING TYPE, THREADS INCLUDED IN SHEAR PLANE

X = BEARING TYPE, THREADS EXCLUDED FROM SHEAR PLANE

* THESE CONDITIONS ARE ALSO APPROPRIATE FOR LONG OR SHORT SLOTTED HOLES NORMAL TO LOAD DIRECTION, WHICH ARE REQUIRED IN BEARING-TYPE CONNECTIONS.

DOUBLE SHEAR CONNECTIONS HAVE TWICE THE VALUE OF SINGLE SHEAR CONNECTIONS.

FOR THREADED PARTS OF ALL OTHER METALS, $F_v = 0.17\ F_u$ FOR "N" TYPE CONNECTIONS, AND $F_v = 0.22\ F_u$ FOR "X" TYPE CONNECTIONS.

FIGURE 3-44a Allowable loads for bolts and threaded parts.

1. Type of steel, A490
2. Type of connection, friction type
3. Bolt diameter, 1 in

The connection in each shear plane is A490 F. In Fig. 3-44a find "A490 F" in the column on the left, then follow the solid line at its right downward into the graph until the line meets the vertical "1-in" diameter line at 17.3 kips. Thus the total allowable shear load is 17.3 × 2 shear planes × 2 bolts = 69.2 kips.

As the allowable shear load of 69.2 kips is less than the bearing load of 110 kips, the shear load governs and the allowable load for the joint is 69.2 kips.

Weld Design: Allowable Load

What is the allowable load per linear inch of a ⅜-in fillet weld made with E70 electrodes? (See Fig. 3-45.)

$$L = 0.707 S f_v$$

L = allowable load per linear inch of weld, ? kips
S = size of weld, ⅜ in
f_v = allowable shear stress for E70 electrode is $0.3 f_t$; f_t of an electrode equals its first two numbers in kips/in^2; thus, the f_t of an E70 electrode = 70 kips/in^2 and its f_v = 0.3 × 70 = 21 kips/in^2

$$L = 0.707 \times 0.375 \times 21$$
$$= 5.57 \text{ kips/lin in}$$

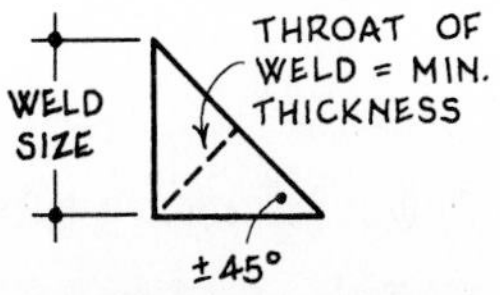

FIGURE 3-45 Weld shape.

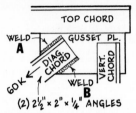

FIGURE 3-46 Welded connection.

Weld Design: Size and Length

A diagonal truss chord of two 2½ x 2 x ¼ in angles is connected to a gusset plate by ³⁄₁₆-in fillet welds as shown in Fig. 3-46. If the maximum stress in the truss chord is 40 kips and the allowable unit shear of the fillet welds is 18 kips/in², how long should be the welds A and B for each angle?

STEP 1. Compute the strength per linear inch of the fillet welds.

$$L = 0.707 S f_v$$

L = strength per linear inch, ? kips
S = size of weld, ³⁄₁₆ in = 0.188 in
f_v = allowable shear stress for the electrode, 18 kips/in²
$L = 0.707 \times 0.188 \times 18$
  $= 2.39$ kips/lin in

STEP 2. Calculate the required total length of the welds.

$$P = LU$$

P = maximum stress in the truss chord, 40 kips
L = required total length of weld, ? in
U = unit strength of weld per linear inch, 2.39 kips

$$40 = L \times 2.39$$
$$L = 16.7 \text{ in}$$

Half the weld length is on one side of the gusset plate, half is on the other side; $16.7/2 = 8.35$ in on each side.

STEP 3. Find the lengths of welds A and B by writing moments about the central axis of the welded angle (Fig. 3-47).

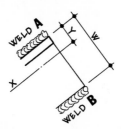

FIGURE 3-47 Detail of weld A and B in Fig. 3-46.

Moment equation about axis X:

$$yA = (W - y)B$$
$$A + B = 8.35 \text{ in}$$
$$A = 8.35 - B$$
$$y(8.35 - B) = (W - y)B$$

y = distance from X axis to weld A (from Table 3-16, y for a 2½ x 2 x ¼ in angle), 0.787 in

B = length of weld B, ? in

W = width of 2½ x 2 x ¼ in angle, 2.5 in

$$0.787(8.35 - B) = (2.5 - 0.787)B$$
$$\text{weld B} = 2.63 \text{ in}$$
$$\text{Weld A} = 8.35 - B = 8.35 - 2.63$$
$$= 5.72 \text{ in}$$

Figure 3.48 is a diagram of common welding symbols.

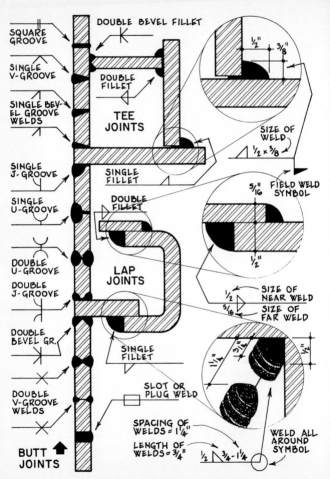

FIGURE 3-48 Welding symbols.

CONCRETE

Table 3-20 lists the allowable stresses for various strengths of concrete.

Shear

A reinforced concrete beam 10 in wide and 24 in deep supports a uniform load of 1600 lb/lin ft on a 27-ft span. What is the total shear? What is the unit shear?

Total shear:

$$V = U - \frac{dW}{L}$$

V = maximum total shear near one end of beam, ? lb

U = maximum shear depending on load arrangement (as listed in Table 3-5, for simple beam, uniformly distributed load), $W/2$

d = depth of concrete beam, 24 in

W = total weight supported by the beam, $1600 \times 27 = 43,200$ lb

L = length of span, 27 ft = 324 in

$$V = \frac{43,200}{2} - \frac{24 \times 43,200}{324}$$
$$= 18,400 \text{ lb}$$

Unit shear:

$$v = \frac{V}{0.85bd_e}$$

v = maximum unit shear near end of beam, ? lb/in^2

V = maximum total shear near end of beam from above, 18,400 lb

b = width of beam, 10 in

d_e = effective depth of beam (from top of beam to center axis of bottom reinforcing, usually 2 in less than total depth), $d - 2 = 24 - 2 = 22$ in

TABLE 3-20 ALLOWABLE CONCRETE STRESSES

Type of stress	Symbol	Allowable stresses, lb/in²			
		$f_c = 2500$	$f_c = 3000$	$f_c = 4000$	$f_c = 5000$
Modulus of elasticity for concrete that weighs 145 lb/ft³ min.	E	2,880,000	3,150,000	3,640,000	4,070,000
Flexure	f_c				
Compression	$0.45f_c$	1125	1350	1800	2250
Tension	$1.6\sqrt{f_c}$	80	88	102	113
Shear	f_v				
Beams with no web reinforcement	$1.1\sqrt{f_c}$	55	60	70	78
Joists with no web reinforcement	$1.2\sqrt{f_c}$	61	66	77	86
Members with stirrups	$5\sqrt{f_c}$	250	274	316	354
Slabs and footings	$2\sqrt{f_c}$	100	110	126	141
Bearing	f_b				
On full area	$0.25f_c$	625	750	1000	1250
On one-third area or less	$0.37f_c$	938	1125	1500	1875
Formula coefficient $f_y = 60$ kips/in²	R	161	204	271	295

$$v = \frac{18,400}{0.85 \times 10 \times 22} = 98.4 \text{ lb/in}^2$$

Bending Moment

A reinforced concrete beam 40 ft long has a dead load of 2300 lb/lin ft and a live load of 3200 lb/lin ft. If $f_c = 3000$ lb/in^2, $f_y = 60,000$ lb/in^2, and $b = 12$ in, design the beam.

STEP 1. Compute the total moment capacity of the beam.

$$M = 1.4M_d + 1.7M_1$$

M = total moment capacity of the beam, ? in·lb

M_d = beam moment capacity due to dead load; assume the concrete beam ends are fixed; from Table 3-5, find the formula for the maximum moment of a beam fixed at both ends with uniform load: $M = WL/12$

W = dead load weight on the beam, $2300 \times 40 = 92,000$ lb

L = length of span, $40 \times 12 = 480$ in

$$M_d = \frac{92,000 \times 480}{12} = 3,680,000 \text{ in·lb}$$

M_1 = beam moment capacity due to live load; use same formula as for dead load: $M = WL/12$

W = live load weight on the beam, $3200 \times 40 = 128,000$ lb

L = length of span, 480 in

$$M_1 = \frac{128,000 \times 480}{12} = 5,120,000 \text{ in·lb}$$
$$M = 1.4 \times 3,680,000 + 1.7 \times 5,120,000$$
$$= 13,900,000 \text{ in·lb}$$

STEP 2. Compute the allowable reinforcement ratio.

$$r = 0.75 \left(\frac{0.72 f_c}{f_y} \times \frac{87,000}{87,000 + f_y} \right)$$

r = maximum allowable reinforcement ratio, ?
f_c = allowable unit stress of concrete, 3000 lb/in^2
f_y = allowable unit stress of reinforcing rods, 60,000 lb/in^2

$$r = 0.75 \left(\frac{0.72 \times 3,000}{60,000} \times \frac{87,000}{87,000 + 60,000} \right)$$
$$= 0.016$$

STEP 3. Calculate the effective depth of the beam.

$$M = 0.9 r f_y b d_e^2 \left(1 - 0.59 r \frac{f_y}{f_c} \right)$$

M = total moment capacity of the beam from above, 13,900,000 in·lb
r = allowable reinforcement ratio from above, 0.016
f_y = allowable unit stress of the rebars, 60,000 lb/in^2
b = width of beam, 12 in
d_e = effective depth of beam (the distance between the top of the beam to the center line of the bottom reinforcing), ? in
f_c = allowable unit stress of the concrete, 3000 lb/in^2

$$13,900,000 = 0.9 \times 0.016 \times 60,000 \times 12$$
$$\times d_e^2 \left(1 - 0.59 \times 0.016 \times \frac{60,000}{3000} \right)$$
$$d_e^2 = 1653$$
$$= 40.7$$

Use 41 in.
    Total depth = $d_e + 2 = 41 + 2 = 43$ in.

STEP 4. Compute the tensile reinforcing area, then select the rebars.

$$A_s = rbd_e$$

A_s = cross-section area of reinforcing steel, ? in^2
r = allowable reinforcement ratio, from step 2, 0.016
b = width of beam, 12 in
d_e = effective depth of beam, 41 in

$$A_s = 0.016 \times 12 \times 41 = 7.87 \text{ in}^2$$

TABLE 3-21 REINFORCING BAR DIMENSIONS

Bar number	Diameter, in	C-S area, in^2	Perimeter, in
3	0.375	0.11	1.18
4	0.500	0.20	1.57
5	0.625	0.31	1.96
6	0.750	0.44	2.36
7	0.875	0.60	2.75
8	1.00	0.79	3.14
9	1.13	1.00	3.54
10	1.27	1.27	3.99
11	1.41	1.56	4.43
14	1.59	2.25	5.32
18	2.25	4.00	7.09

Select the rebars. From Table 3-21, pick a size and number of rebars whose cross-sectional area equals at least 7.87 in^2. Usually it is best to pick an even number of bars.

Use four #18 bars:

$$4 \times 2.25 = 9 \text{ in}^2 > 7.87 \text{ in}^2$$

NOTE: A concrete beam's length of span should not exceed 50 times its width. In continuous concrete beams, certain moment coefficients are permitted by the American Concrete Institute (ACI) Code. Positive

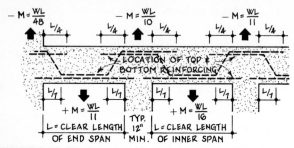

FIGURE 3-49 Bending moment in continuous concrete beams: maximum moments, rebar arrangements, uniform loading.

moments occur between the supports, and negative moments occur over the supports. These moments are illustrated in Fig. 3-49.

Compressive Reinforcement

Compressive reinforcement accomplishes three things: (1) reduces flexural compression stress, (2) reduces long-time deflection, and (3) reduces the possibility of sudden failure.

All compression reinforcement in beams must be held in place by ties or stirrups to prevent buckling of the compression rebars.

Example A 12-in wide by 39-in deep concrete beam contains 9.36 in² of tensile reinforcing. If compressive reinforcing is required in this beam, how much should it have?

$$2 - 1.2\left(\frac{A_c}{A_t}\right) \geq 0.6$$

A_c = cross-sectional area of compressive rebars, ? in²
A_t = cross-sectional area of tensile rebars, 9.36 in²

$$2 - 1.2\left(\frac{A_c}{9.36}\right) \geq 0.6$$

$$A_c = 10.9 \text{ in}^2 \text{ minimum}$$

Diagonal Tension: Stirrups

A reinforced concrete beam has a span of 28 ft and supports a uniformly distributed total load of 122,000 lb. If $f_c = 3000$ lb/in^2, $b = 11$ in, $d_e = 24$ in, and tensile reinforcing = three #11 bars, design the stirrups if required.

STEP 1. Compute the beam's unit shear stress. At $f_c = 3000$ lb/in^2, if v is greater than 60 lb/in^2, stirrups are required.

$$v = \frac{W/2 - (dW/L)}{0.85bd_e}$$

v = actual unit shear stress at beam end, ? lb/in^2
W = total weight supported by the beam, 122,000 lb
d = depth of beam: $d_e = 24$ in, $d = d_e + 2 = 24 + 2 = 26$ in
L = length of beam, 28 ft = $28 \times 12 = 336$ in
b = width of beam, 11 in
d_e = effective depth of beam, 24 in

$$v = \frac{122,000/2 - (26 \times 122,000/336)}{0.85 \times 11 \times 24} =$$
$$= 230 \text{ lb/in}^2 \qquad > 60 \text{ lb/in}^2$$

Stirrups are required.

STEP 2. Design the stirrups. The maximum spacing is $d_e/2$, each stirrup must be at least #3 rebar, and the cross-sectional area of each stirrup must be at least $0.0015bs$.

Spacing:

$$s \le \frac{d_e}{2} \qquad d_e = 24 \text{ in max}$$
$$s \le 24/2 = 12 \text{ in maximum spacing}$$

Cross-sectional area:

$$A_s \ge 0.0015bs$$

A_s = cross-sectional area of each stirrup, ? in^2
b = width of beam, 11 in
s = stirrup spacing from above, 12 in

$$A_s \geq 0.0015 \times 11 \times 12 = 0.198 \text{ in}^2$$

Use #4 rebar for stirrups:

$$A_s = 0.20 \text{ in}^2 > 0.198 \text{ in}^2$$

Deflection

A simply supported reinforced concrete beam is 36 ft long and supports a dead load of 1750 lb/lin ft and a live load of 1400 lb/lin ft. If $f_c = 4000$ lb/in^2, $b = 12$ in, and $d = 28$ in, what are the beam's immediate and long-term deflections? There is no compressive reinforcement.

STEP 1. Calculate the immediate deflection. Use the formula from Table 3-5 for a simple beam with uniform load.

$$\Delta = \frac{5WL^3}{384EI}$$

Δ = immediate deflection of the beam, ? in
W = weight of total load on the beam, $36(1750 + 1400) = 113,000$ lb
L = length of span, 36 ft = $36 \times 12 = 432$ in
E = modulus of elasticity of the beam (from Table 3-20, when $f_c = 4000$ lb/in^2), 3,640,000 lb/in^2
I = moment of inertia of beam section; from Table 3-9, I of a rectangle with axis through center = $BD^3/12$:
  B = width of beam, 12 in
  D = depth of beam, 28 in

$$\frac{BD^3}{12} = \frac{12 \times 28^3}{12} = 21,900 \text{ in}^4$$

$$\Delta = \frac{113,400 \times 432^3}{384 \times 3,640,000 \times 21,900} = 0.298 \text{ in}$$

STEP 2. Calculate the long-term deflection.

$$\Delta_l = \Delta \left(\frac{DL}{LL + DL} \right) \left(2 - 1.2 \frac{A_c}{A_t} \right)$$

Δ_l = long-term deflection, ? in

Δ = immediate deflection, 0.298 in

DL = dead load per linear foot, 1750 lb

LL = live load per linear foot, 1400 lb

A_c = cross-sectional area of compressive rebar, 0 in^2

A_t = cross-sectional area of tensile rebars; if A_c is zero, amount of A_t is irrelevant

$$\Delta_l = 0.298 \left(\frac{1750}{1400 + 1750} \right) (2 - 0)$$
$$= 0.331 \text{ in}$$

Total long-term deflection = $0.298 + 0.331 = 0.629$ in.

Are these deflections safe?

Immediate deflection:

$$\Delta \left(\frac{LL}{LL + DL} \right) \leq \frac{L}{360}$$
$$0.298 \left(\frac{1400}{1400 + 1750} \right) \leq \frac{432}{360}$$
$$0.132 \leq 1.2 \text{ in} \qquad \text{OK}$$

Long-term deflection:

$$\Delta_l \leq \frac{L}{240}$$
$$0.629 \leq \frac{432}{240}$$
$$0.629 \leq 1.8 \text{ in} \qquad \text{OK}$$

Moment of Inertia of Irregular Cross Section

A reinforced concrete mezzanine bridge overlooking a large hotel lobby
has a cross section as shown in Fig. 3-50. What is the moment of inertia
of this construction?

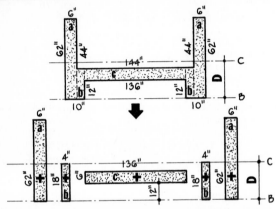

FIGURE 3-50 Concrete cross section 1.

STEP 1. Locate the horizontal central axis through the cross section
by taking moments about the base after dividing the total cross-section
area into component rectangles.

$$A_a D_a + A_b D_b + A_c D_c = A_t D$$

A_a = cross-sectional area of rectangles a, $2 \times 6 \times 62 = 744$ in^2

D_a = moment arm distance from center of rectangles a to base, $D_a = 62/2$
   = 31 in

A_b = cross-section area of rectangles b, $2 \times 4 \times 18 = 144$ in^2

D_b = moment arm distance from center of rectangles b to base, $D_b = 18/2$
   = 9 in

A_c = cross-section area of rectangle c, $6 \times 136 = 816 \text{ in}^2$

D_c = moment arm distance from center of rectangle c to base, $12 + 6/2 = 15 \text{ in}$

A_t = total area of cross section, $A_a + A_b + A_c = 744 + 144 + 816 = 1704 \text{ in}^2$

D = moment arm distance from central axis to base, ? in

$$744 \times 31 + 144 \times 9 + 816 \times 15 = 1704D$$
$$D = 21.5 \text{ in}$$

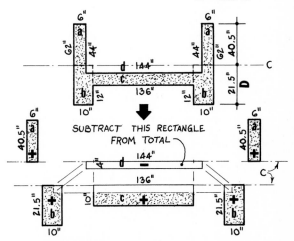

FIGURE 3-51 Concrete cross section 2.

STEP 2. Calculate the total moment of inertia by dividing the cross section into smaller areas and then computing I for each rectangle as shown in Fig. 3-51. From Table 3-9, I for rectangle with axis through base = $bd^3/3$.

$$I_t = \frac{b_a d_a^3}{3} + \frac{b_b d_b^3}{3} + \frac{b_c d_c^3}{3} - \frac{b_d d_d^3}{3}$$

I_t = moment of inertia of total cross section, ? in^4
b_a = base of rectangles a, $6 \times 2 = 12$ in
d_a = depth of rectangles a, 40.5 in
b_b = base of rectangles b, $10 \times 2 = 20$ in
d_b = depth of rectangles b, 21.5 in
b_c = base of rectangle c = 136 in
d_c = depth of rectangle c, 10 in
b_d = base of rectangle d, 144 in
d_d = depth of rectangle d, 4 in

$$I_t = \frac{12 \times 40.5^3}{3} + \frac{20 \times 21.5^3}{3} + \frac{136 \times 10^3}{3} - \frac{144 \times 4^3}{3}$$
$$= 374,000 \text{ in}^4$$

One-Way Slab Design

This system is economical for medium and heavy live loads on spans ranging from 6 to 12 ft. The slab is usually integral with reinforced concrete beams designed as T beams.

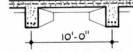

$\overline{\quad\quad\text{IO}'\text{-}\text{O}''\quad\quad}$

FIGURE 3-52 Concrete one-way slab.

Example A one-way concrete slab is poured with beams 10 ft apart as shown in Fig. 3-52. Live load = 100 lb/ft^2, dead load = 75 lb/ft^2, f_y = 60,000 lb/in^2, f_c = 4000 lb/in^2. Design the slab thickness and reinforcing.

STEP 1.   Determine the slab thickness. For one-way slabs:

$t \geq L/25$ for simple spans
$t \geq L/30$ for one-end continuous spans

$$t \geq L/35 \text{ for both ends continuous spans}$$
$$t = 4 \text{ in minimum}$$

Since both ends are continuous, $t \geq L/35$.

$$L = 10 \text{ ft} = 120 \text{ in}$$
$$t \geq \frac{120}{35} = 3.43 \text{ in}$$

Use 4 in.

STEP 2.   Design the tensile reinforcing. Compute the reinforcement ratio, as in "Bending Moment," step 2, page 170.

$$r = 0.75 \left(\frac{0.72 f_c}{f_y} \right) \left(\frac{87,000}{87,000 + f_y} \right) = 0.0213$$

Find the effective depth.

$$d_e = d - 2 = 4 - 2 = 2 \text{ in}$$

Compute the tensile rebar area, as in "Bending Moment," step 4, page 170.

$$A_s \geq rbd_e = 0.511 \text{ in}^2$$

Select the rebar from Table 3-21. Use #5 rebars at 7 in o.c.:

$$0.31 \times \frac{12}{7} = 0.531 \text{ in}^2 > 0.511 \text{ in}^2$$

STEP 3.   Determine the amount of temperature reinforcing. This is laid perpendicular to the tensile rebars. Temperature rebars should be placed no farther apart than 18 in or 5 times the slab thickness, whichever is less.

Area of temperature rebars:

$$A_t \geq 0.0018 bd$$

$b = 12$ in
$d = 4$ in

$$A_t \geq 0.0018 \times 12 \times 4 = 0.0864 \text{ in}^2$$

Use #3 rebars at 15 in o.c.:

$$0.11 \times \frac{12}{15} = 0.088 \text{ in}^2 > 0.0864 \text{ in}^2$$

T Beam Design

48-ft long reinforced concrete T beams poured integrally with a 4-in concrete slab on top are spaced 10 ft apart. Live load = 100 lb/ft^2, dead load = 75 lb/ft^2, $f_y = 60{,}000$ lb/in^2, $f_c = 4000$ lb/in^2. Design the T beam depth, width, and reinforcing.

STEP 1. Calculate the maximum negative (over the support) moment capacity. From Fig. 3-49, maximum negative moment = $WL/11$.

$$-M = 1.4M_d + 1.7M_l \qquad M_d \text{ and } M_l = \frac{WL}{11}$$

$$= 1.4\left(\frac{W_d L}{11}\right) + 1.7\left(\frac{W_l L}{11}\right)$$

$-M$ = total maximum moment capacity over the support, ? in·lb
W_d = weight of dead load, $75 \times 48 \times 10 = 36{,}000$ lb
W_l = weight of live load, $100 \times 48 \times 10 = 48{,}000$ lb
L = length of span, 48 ft = 576 in

$$-M = 1.4\left(\frac{36{,}000 \times 576}{11}\right) + 1.7\left(\frac{48{,}000 \times 576}{11}\right)$$
$$= 6{,}910{,}000 \text{ in·lb}$$

STEP 2. Find the depth of beam at negative moment.

$$d_e = \sqrt{\frac{-M}{Rb}}$$

d_e = effective depth of beam, ? in
M = maximum moment, 6,910,000 in·lb
R = formula coefficient (from Table 3-17, R for f_c), 4000 lb/in^2 = 271.
b = width of T beam base: $b \approx L/36$, therefore $576/36 \approx 16$ in

$$d_e = \sqrt{\frac{6,910,000}{271 \times 16}} = 39.9 \text{ in}$$

Use 40 in.
   Total depth:

$$d_e + 2 = 40 + 2 = 42 \text{ in}$$

STEP 3.   Check for unit shear. Compute according to step 1 under "Diagonal Tension: Stirrups," page 173. $W = 84,000$ lb, $d = 42$ in, $L = 516$ in, $b = 16$ in, $d_e = 40$ in.

$$v = \frac{W/2 - (dW/L)}{0.85bd_e} = 65.9 \text{ lb/in}^2$$

   When $f_c = 4000$ lb/in^2, no stirrups are required if v is less than 70 lb/in^2.

STEP 4.   Compute the maximum positive moment at center of span. From Fig. 3-49, maximum positive moment = $WL/16$.

$$M = 1.4M_d + 1.7M_l \qquad M_d \text{ and } M_l = \frac{WL}{16}$$
$$= 1.4\left(\frac{W_dL}{16}\right) + 1.7\left(\frac{W_lL}{16}\right)$$

$W_d = 36,000$ lb, $W_l = 48,000$ lb, $L = 576$ in

$$M = 1.4 \left(\frac{36,000 \times 576}{16} \right) + 1.7 \left(\frac{48,000 \times 576}{16} \right)$$
$$= 4,750,000 \text{ in·lb}$$

STEP 5. Determine the amount of tensile reinforcing at midspan.

$$A_s \geq \frac{M}{0.85 f_y [d_e - (t/2)]}$$

A_s = cross-sectional area of tensile reinforcing, ? in^2
M = positive moment capacity from above, 4,750,000 in·lb
f_y = unit stress of reinforcing steel, 60,000 lb/in^2
d_e = effective depth of beam from step 2, 40 in
 t = thickness of floor slab, 4 in

$$A_s \geq \frac{4,750,000}{0.85 \times 60,000 [40 - (4/2)]} = 2.45 \text{ in}^2$$

Use four #7 rebars:

$$0.60 \times 4 = 2.40 \text{ in}^2 \qquad \text{Close enough}$$

STEP 6. Determine the amount of reinforcing over the the supports.

$$A_s \geq \frac{-M}{0.35 f_y d_e}$$

A_s = cross-sectional area of tensile reinforcing, ? in^2
M = negative moment from step 1, 6,910,000 in·lb
f_y = unit stress of reinforcing steel, 60,000 lb/in^2
d_e = effective depth of beam from step 2, 40 in

$$A_s \geq \frac{6,910,000}{0.35 \times 60,000 \times 40} = 8.23 \text{ in}^2$$

Use four #14 rebars:

$$2.25 \times 4 = 9.00 \text{ in}^2 > 8.23 \text{ in}^2$$

Ribbed Slab Construction Design

Commonly known as *pan joist construction,* this system is suitable for fairly long spans that have light to medium live loads. Standard pan forms run 6 to 20 in deep and 20 or 30 in wide; the top slab is usually 2½ to 4 in thick.

Because pan joists resemble small adjacent T beams, their design is similar to T beam design.

Example A reinforced concrete pan joist system has a clear span of 24 ft, ribs 6 in wide and pans 30 in wide at the bottom, and a top slab 3 in thick. Live load = 60 lb/ft², dead load = 90 lb/ft², f_y = 60,000 lb/in², f_c = 4000 lb/in². Design the depth and reinforcing. (See Fig. 3-53.)

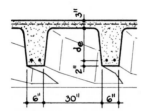

FIGURE 3-53 Concrete ribbed slab.

STEP 1. Calculate the negative moment. Compute according to step 1 under "T Beam Design" above.

$$-M = 1.4 \left(\frac{W_d L}{11} \right) + 1.7 \left(\frac{W_l L}{11} \right)$$

$-M$ = maximum negative moment over the support, ? in·lb
W_d = weight of dead load, 90 × 3 × 24 = 6480 lb
W_l = weight of live load, 60 × 3 × 24 = 4320 lb
L = length of span, 24 ft = 288 in

$$-M = 1.4 \left(\frac{6{,}480 \times 288}{11} \right) + 1.7 \left(\frac{4{,}320 \times 288}{11} \right)$$
$$= 430{,}000 \text{ in} \cdot \text{lb}$$

STEP 2.   Find the depth of joist at negative moment.

$$d_e = \sqrt{\frac{-M}{Rb}}$$

d_e = effective depth of joist, ? in

$-M$ = maximum negative moment from above, 430,000 in·lb

R = formula coefficient (from Table 3-17, R for f_c), 4000 lb/in² = 271

b = average width of joist (bottom is 6 in; because pans taper, add 1 in for average width), 7 in

$$d_e = \sqrt{\frac{430{,}000}{271 \times 7}}$$
$$= 15.1$$

Use 15 in.

STEP 3.   Check for unit shear.

$$v = \frac{W/2 - (dW/L)}{0.85bd_e}$$

v = actual unit shear (if it exceeds 70 lb/in², the allowable unit stress for beams with no web reinforcement at f_c = 4000 lb/in², from Table 3-17, tapered end pans are required), ? lb/in²

W = weight of load on joist (live load + dead load), 6,480 + 4,320 = 10,800 lb

d = total depth of beam, d_e + 2 = 15 + 2 = 17 in

L = length of span, 288 in

b = average width of joist, from step 1, 7 in

d_e = effective depth of beam, from step 1, 15 in

$$v = \frac{10,800/2 - (17 \times 10,800/288)}{0.85 \times 7 \times 15} = 53.4 \text{ lb/in}^2$$

This is less than 70 lb/in², therefore no tapered end pans are required.

If tapered end pans are required, add 4 in (the amount the pans increase each joist's width at its ends) to b in the above formula and recalculate v. If v is still greater than the allowable unit sheer stress, the depth of the joists must be increased.

STEP 4.   Compute the maximum positive moment at midspan.

$$M = 1.4 \left(\frac{W_d L}{16}\right) + 1.7 \left(\frac{W_l L}{16}\right)$$

$W_d = 6480$ lb, $W_l = 4320$ lb, $L = 288$ in.

$$M = 1.4 \left(\frac{6,480 \times 288}{16}\right) + 1.7 \left(\frac{4,320 \times 288}{16}\right)$$

$$= 295,000 \text{ in} \cdot \text{lb}$$

STEP 5.   Determine the amount of tensile reinforcing at midspan.

$$A_s \geq \frac{M}{0.85 f_y [d_e - (t/2)]}$$

A_s = cross-sectional area of tensile reinforcing, ? in²
M = positive moment from above, 295,000 in·lb
f_y = unit stress of reinforcing steel, 60,000 lb/in²
d_e = effective depth of joist from step 2, 15 in
t = thickness of floor slab, 3 in

$$A_s \geq \frac{295,000}{0.85 \times 60,000 \, [15 - (3/2)]} = 0.428 \text{ in}^2$$

Use one #6 rebar:

$$A_s = 0.44 \text{ in}^2 > 0.428 \text{ in}^2$$

STEP 6. Determine the amount of tensile reinforcing over the supports.

$$A_s \geq \frac{-M}{0.35 f_y d_e}$$

A_s = cross-sectional area of tensile reinforcing, ? in^2
$-M$ = 430,000 in·lb, f_y = 60,000 lb/in^2, d_e = 15 in

$$A_s \geq \frac{430,000}{0.35 \times 60,000 \times 15} = 1.37 \text{ in}^2$$

Use one #11 rebar:

$$A_s = 1.56 \text{ in}^2 > 1.37 \text{ in}^2$$

STEP 7. Determine the amount of temperature reinforcing, to be placed perpendicular to joists in floor slab. Maximum spacing is 18 in or 5 times the slab thickness, whichever is less.

$$A_t \geq 0.0018bt$$

A_t = cross-sectional area of temperature reinforcing, ? in^2
b = width of a unit section of flooring (use 1 ft wide strip), 12 in
t = thickness of floor slab, 3 in

$$A_t \geq 0.0018 \times 12 \times 3$$
$$= 0.0648 \text{ in}^2/\text{lin ft}$$

As this area is much less than the smallest rebar area (A_s = 0.11 in^2), consider using welded wire fabric. From Table 3-22, select a wire size and spacing that will satisfy the above A_s requirements. From inspection, the best choice is W5 wire at 9 in o.c.

W5 wire at 9 in o.c.:

$$A_s = 0.67 \text{ in}^2/\text{lin ft} > 0.648 \text{ in}^2/\text{lin ft}$$

TABLE 3-22 PROPERTIES OF WELDED WIRE FABRIC

Smooth	Deformed	Diameter, in	Cross-sectional area per linear foot		
			6 in o.c.	9 in o.c.	12 in o.c.
W2	—	0.159	0.04	0.027	0.02
W3	—	0.195	0.06	0.04	0.03
W4	D4	0.225	0.08	0.053	0.04
W5	D5	0.252	0.10	0.067	0.05
W6	D6	0.276	0.12	0.08	0.06
W7	D7	0.298	0.14	0.093	0.07
W8	D8	0.319	0.16	0.107	0.08
W9	D9	0.338	0.18	0.12	0.09
W10	D10	0.356	0.20	0.133	0.10
W11	D11	0.374*	0.22	0.147	0.11

*Diameter approximately equal to #3 rebar.

Data taken from the *Welded Wire Fabric Manual of Standard Practice* with the permission of The Wire Reinforcement Institute.

Concrete Staircases

A four-story motel has reinforced concrete stairs. Each 10-ft 6-in flight has two 5-ft 3-in tall staircases with halfway landings. If the steps are 4 ft 8 in wide, the treads are 11 in, the risers are 7 in, and the local building code lists the allowable live load for stairs at 100 lb/ft^2, how should the stairs be designed?

$$f_c = 4000 \text{ lb/in}^2$$
$$f_y = 60,000 \text{ lb/in}^2$$

STRATEGY A staircase is basically an inclined beam whose span is in the horizontal plane, as shown in Fig. 3-54b.

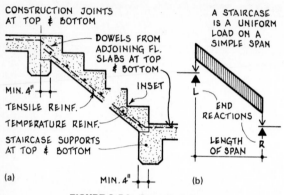

FIGURE 3-54 Concrete staircase.

STEP 1. Determine the horizontal distance between the staircase supports.

$$L = \frac{Ht}{12rN}$$

L = horizontal length of span between staircase supports, ? ft
H = floor-to-floor height between top finished floor and bottom finished floor of staircase, 10 ft 6 in = $10 \times 12 + 6$ = 126 in
t = tread length, 11 in
r = riser height, 7 in
N = number of equal sections total staircase is divided into, 2 equal sections

$$L = \frac{126 \times 11}{12 \times 7 \times 2} = 8.25 \text{ ft horizontal dimension}$$

STEP 2. Determine the thickness of slab, tensile reinforcing, and temperature reinforcing from Table 3-23.

TABLE 3-23 REINFORCED CONCRETE STAIRCASE DIMENSIONS

Horizontal span, ft	Thickness of slab at inset, in	Tensile reinforcing, in o.c.	Temperature reinforcing, in o.c.
5	4	#3 at 8½	#3 at 18
6	4	#3 at 6	#3 at 18
7	4½	#4 at 9	#3 at 18
8	4½	#4 at 7	#3 at 17
9	5	#4 at 6	#3 at 15
10	5½	#4 at 5½	#3 at 13½
11	6	#5 at 8	#3 at 12
12	6½	#5 at 7½	#3 at 11
13	7	#5 at 7	#3 at 10
14	7½	#5 at 6½	#3 at 9
15	8	#5 at 6	#3 at 8

f_c = 3000 lb/in^2; f_y = 60,000 lb/in^2; LL = 100 lb/ft^2; W_{conc} = 145 lb/ft^3

From examination of Table 3-23:

Horizontal length of slab, 8.25 ft (use 9 ft)
Thickness of slab, 5 in at inset
Tensile reinforcing, #4 at 6 in o.c.
Temperature reinforcing, #3 at 15 in o.c.

Flat-Slab Floors

A flat-slab floor system has a reinforced concrete floor slab having no beams or girders and resting on columns that form square or near-square bays. Between the slabs and columns are square areas of increased thickness, called *drop panels,* and flared column heads. The slab reinforcing consists of two sets of rebars laid at right angles to each other. This system offers simple and duplicatable formwork, lower floor-to-floor heights, and ease of installing mechanical equipment. Flat-slab floors are suitable for warehouses, industrial buildings, and other archi-

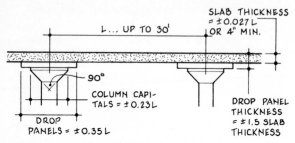

FIGURE 3-55   Concrete flat slab.

tecture having rough finish characteristics and heavy live loads. (See Fig. 3-55.)

The mathematics involved in designing flat-slab floors are too complex to be described here. Instead, a few dimensioning guidelines for designing this system are given below.

1. The live and dead loads should be uniformly distributed.
2. The bays should be square or near-square, up to 30 ft for the longer span.
3. At least three bays should run in each direction, with the longest and shortest span lengths varying by no more than 20 percent.
4. The cantilever around the outer columns should be 0.17 to 0.21 the longer span.
5. The drop panel side (or diameter, if circular) should be 0.33 to 0.37 the longer span.
6. The tops of the column capitals should be 0.20 to 0.25 the longer span.
7. The capital cone should have a 90° vertex angle.
8. The slab thickness should be 0.025 to 0.028 the longer span, but never less than 4 in.
9. The drop panel thickness should be 1.5 the slab thickness.

Flat-Plate Floors

A flat-plate floor is simply a flat floor slab supported by round or square concrete columns. It has much in common with flat-slab systems. It is simpler to build than flat-slab construction but weighs more concrete per square foot of area because a thicker floor slab is required.

Figure 3-56 illustrates the relationship between a flat-plate system's live load, longer span, and slab thickness. Service load = live load + dead load.

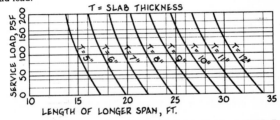

FIGURE 3-56 Flat-plate floor thickness graph for f_c = 3000 lb/ft². *(Reprinted with permission of the Portland Cement Association.)*

Columns: Square or Rectangular Cross Sections

In square or rectangular columns, the minimum thickness is 8 in, the minimum rebar size for vertical steel is #5, the minimum number of bars is four, every corner and alternate bar must be tied in both directions, the maximum distance between bars is 6 in, and the outer layer of concrete over the reinforcing must be at least 1½ in thick.

Example A square reinforced concrete column supports 72 kips live load and 85 kips dead load. If the clear span height of the column is 9 ft 6 in, what should be its cross-sectional area and reinforcing? f_c = 5000 lb/in². f_y = 60,000 lb/in².

STEP 1. Compute the design load from the formula below. (If design loads were used to size the structural members resting on the column, omit this step by using the structural members' total design load for P.)

$$P_d = 1.4\text{DL} + 1.7\text{LL}$$

P_d = total design load on the column, ? kips
DL = dead load on the column, 85 kips
LL = live load on the column, 72 kips

$$P_d = 1.4 \times 85 + 1.7 \times 72 = 241 \text{ kips}$$

STEP 2. Multiply the design load by the *capacity reduction and eccentricity factor* for rectangular columns = 1.9.

$$241 \times 1.9 = 458 \text{ kips} = P_e, \text{ the effective load}$$

STEP 3. Try a column size. The problem is trial and error from here on. Try 10 x 10 in.

STEP 4. Reduce the allowable concrete stress by the slenderness ratio.

$$S = 1.07 - \frac{27h}{100t}$$

S = slenderness ratio for the column, ?
h = clear span height of the column, 9 ft 6 in = 114 in
t = minimum thickness of column, 10 in

$$S = 1.07 - \frac{27 \times 114}{1000 \times 10}$$
$$= 0.762$$

STEP 5. Calculate the vertical rebar size.

$$P_e = 0.85Sf_cA + A_s(f_y - 0.85Sf_c)$$

P_e = effective load on the column from step 2, 458 kips
S = slenderness ratio from step 4, 0.762

f_c = allowable concrete unit stress, 5000 lb/in² = 5 kips/in²
A = total cross-sectional area of column, 10 × 10 = 100 in²
A_s = cross-sectional area of reinforcing steel, ? in²
f_y = allowable reinforcing steel unit stress, 60,000 lb/in² = 60 kips/in²

$$458 = 0.85 \times 0.762 \times 5 \times 100 + A_s(60 - 0.85 \times 0.762 \times 5)$$
$$A_s = 2.38 \text{ in}^2$$

STEP 6. Check to see that A_s is between 0.01 and 0.08A. If A_s is less than 0.01A, either increase the size of rebar or go back to step 3 and try a smaller column size. If A_s is more than 0.08A, go back to step 3 and try a larger column size.

$$\frac{A_s}{A} = \frac{2.38}{100} = 0.0238 \quad \text{OK}$$

10 x 10 is OK.

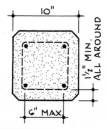

FIGURE 3-57 Column cross section.

STEP 7. Design the vertical reinforcing (see Fig. 3-57). The maximum distance between rebars is 6 in o.c., and minimum thickness of the outer concrete is 1.5 in. Use four rebars and locate to satisfy these conditions.

$$A_s \geq 2.38 \text{ in}^2$$

Use four #7 rebars:

$$4 \times 0.60 = 2.40 \text{ in}^2 > 2.38 \text{ in}^2$$

STEP 8.   Design the tie bar size and spacing.
Size:

#3 if vertical rebars are #10 or less
#4 if vertical rebars are #11 or more

Vertical rebars are #7. Use #3 rebars.
   Spacing: Use the least distance of a, b, and c:

a. Spacing no greater than 16 vertical rebar diameters; diameter of #5 rebar = 0.625 in: $0.625 \times 16 = 10$ in
b. Spacing no greater than 48 tie bar diameters; diameter of #3 rebar = 0.375 in: $0.375 \times 48 = 18$ in
c. Spacing no greater than the minimum thickness of column; minimum thickness of $10 \times 10 = 10$ in +

The least distance is 10 in.
   Tie bars: Use #3 rebars at 10 in o.c.

Columns with Round Cross Sections

In round columns, the minimum thickness is 10 in, the minimum rebar size for vertical steel is #5, the minimum number of bars is six, the clear spacing between spirals of the lateral reinforcing is at least 1 in and not more than 3 in, the minimum size of the spiral rebar is #3, and f_y cannot be more than 60,000 lb/in².

Example   A 12-ft tall round column has a design load of 1600 kips. Size the column and reinforcing. f_c = 4000 lb/in². f_y = 60,000 lb/in².

STEP 1.   Compute the design load from the formula below. (If design loads were used to size the structural members resting on the column, omit this step by using the structural member's total design load for P.)

$P_d = 1.4DL + 1.7LL$ (see "Columns: Square or Rectangular Cross Sections" step 1, page 192.)

Omit this step, as the load given is the design load.

STEP 2.   Multiply the design load by the *capacity reduction and eccentricity factor* for round columns = 1.6.

$$1600 \times 1.6 = 2{,}560 \text{ kips} = P_e \text{, the effective load}$$

STEP 3.   Try a column diameter. The problem is trial and error from here on. Try a 24-in diameter.

STEP 4.   Reduce the allowable concrete unit stress by the slenderness ratio.

$$S = 1.07 - \frac{27h}{1000d}$$

S = slenderness ratio for the column, ?
h = clear span height of the column, 12 ft = 144 in
d = diameter of the column, 24 in

$$S = 1.07 - \frac{27 \times 144}{1000 \times 24}$$
$$= 0.908$$

STEP 5.   Calculate the vertical rebar cross-sectional area.

$$P_e = 0.85Sf_cA + A_s(f_y - 0.85Sf_c)$$

P_e = effective load on the column from step 2, 2560 kips
S = column slenderness ratio from step 4, 0.908
f_c = allowable unit stress for concrete, 4000 lb/in^2 = 4 kips/in^2
A = total cross-sectional area of the column, $0.785d^2 = 0.785 \times 24^2$
    = 452 in^2
A_s = vertical rebar cross-sectional area, ? in^2
f_y = allowable unit stress for rebars, 60,000 lb/in^2 = 60 kips/in^2

$$2560 = 0.85 \times 0.908 \times 4 \times 452 + A_s(60 - 0.85 \times 0.91 \times 4)$$
$$A_s = 20.4 \text{ in}^2$$

STEP 6. Check to see that A_s is between 0.01 and $0.08A$. If A_s is less than $0.01A$, either use larger rebars or go back to step 3 and try a smaller column. If A_s is greater than $0.08A$, go back to step 3 and try a larger column.

$$\frac{A_s}{A} = \frac{20.4}{452} = 0.0453 \qquad \text{OK}$$

24-in diameter is OK.

STEP 7. Design the vertical reinforcing. The minimum number of rebars is six.

$$A_s \geq 20.4 \text{ in}^2$$

Use sixteen #10 rebars:

$$16 \times 1.27 = 20.32 \text{ in}^2 \qquad \text{Close enough}$$

STEP 8. Determine the amount of spiral reinforcing.

$$S = \frac{f_y A_s(d - 4 - D_s)}{0.89 f_c(d - 2)}$$

S = pitch per revolution of spiral reinforcing, ? in per revolution
f_y = allowable unit stress of reinforcing steel, 60,000 lb/in^2 = 60 kips/in^2
A_s = cross-sectional area of spiral reinforcing (use #3 bar), 0.11 in^2
d = diameter of column, 24 in
D_s = diameter of spiral reinforcing bars (diameter of #3 bar), 0.375 in
f_c = allowable unit stress for concrete, 4000 lb/in^2 = 4 kips/in^2

$$S = \frac{60 \times 0.11(24 - 4 - 0.375)}{0.89 \times 4(24 - 2)} = 1.65 \text{ in}$$
$$= 1.65 \text{ in} \qquad \text{Use } 1\tfrac{5}{8} \text{ in.}$$

Spiral reinforcing: #3 rebars at 1⅝-in pitch per revolution.

Bearing Walls

Reinforced concrete bearing walls must be at least 6 in thick, the ratio of unsupported height to wall thickness must not exceed 25, and the load centroids resting on the wall should bear within the middle third of its thickness. Basement, fire, party, and foundation walls must be at least 8 in thick.

Example A 13-ft 4-in tall reinforced concrete bearing wall supports a roof made of 5-ft wide concrete double-T beams that cantilever 4 ft beyond the outer face of the wall. Each double-T beam supports a live load of 8400 lb and a dead load of 10,600 lb; the T stems are 3¾ in wide at the bottom and 2 ft 6 in apart. How thick should the bearing wall be and how much reinforcing should it have? $f_c = 3000$ lb/in². $f_y = 60,000$ lb/in². (See Fig. 3-58.)

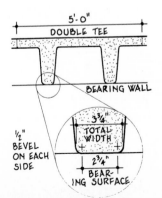

FIGURE 3-58 Double T on bearing wall.

STEP 1. Compute the design load from the formula below. (If design loads were used to size the structural members resting on the bearing wall, use them and omit this step.)

$$P_d = 1.4LL + 1.7DL$$

P_d = design load, ? kips
LL = live load on the wall, 8400 lb per double T = 8.4 kips
DL = dead load on the wall, 10,600 lb per double T = 10.6 kips

$$P_d = 1.4 \times 8.4 + 1.7 \times 10.6 = 29.8 \text{ kips}$$

STEP 2. Try a wall thickness, then later check it with reference to load criteria. Pick a thickness (t) that is approximately one eighteenth the height.

$$t \approx \frac{h}{18} \approx \frac{160}{18} \approx 9 \text{ in.}$$

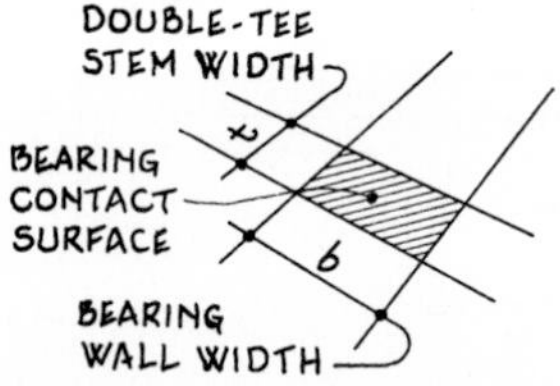

DOUBLE-TEE
STEM WIDTH

BEARING
CONTACT
SURFACE

BEARING
WALL WIDTH

FIGURE 3-59 Bearing contact surface.

STEP 3. Check t according to bearing contact surface between double-T stems and top of wall. (See Fig. 3-59.)

$$t = \frac{P_d}{0.55bf_b}$$

t = trial thickness of wall, 8 in

P_d = design load on wall, 29.8 kips per double T; half that much per stem, 29.8/2 = 14.9 kips

b = width of T stem; total width = 3¾ = 3.75 in, but each bottom edge of a T stem has a ½-in bevel, thus the actual bearing surface of each stem is 3.75 − 2 × 0.5 = 2.75 in

f_b = allowable concrete unit bearing stress, from Table 3-20, f_b = 0.37
     f_c = 1110 lb/in² = 1.11 kips/in²

$$t = \frac{14.9}{0.55 \times 2.75 \times 1.11} = 8.87 \qquad \text{less than 9 in} \qquad \text{OK}$$

STEP 4. Find the effective bearing length of the wall. This dimension is found according to the two criteria below. The least value governs.

L = center-to-center dimension of concentrated loads on the bearing wall. The double-T stems are spaced at 2 ft 6 in o.c. Thus, L = 2 ft 6 in = 30 in.

L = $b + 4t$; from above, b = 2.75 in and t = 8 in. Thus, L = 2.75 + 4 × 8 = 34.75 in.

The least value is 30 in.

STEP 5. Check t according to the slenderness ratio of the wall.

$$P_d \leq 0.385 f_b L t \left[1 - \left(\frac{H}{40t} \right)^2 \right]$$

P_d = design load on the bearing wall from step 1, 29.8 kips

f_b = allowable concrete unit bearing stress from Table 3-20, f_b = 0.25
     f_c = 750 lb/in² = 0.75 kips/in²

L = effective bearing length of the wall from step 4, 30 in

t = trial thickness of wall. 9 in

H = unsupported height of the bearing wall, 160 in

$$29.8 \leq 0.385 \times 0.75 \times 30 \times 8 \left[1 - \left(\frac{160}{40 \times 9} \right)^2 \right]$$

$$29.8 \leq 55.6$$

9-in wall thickness is OK.

STEP 6. Size the vertical and horizontal reinforcing according to the formulas below. Maximum spacing of rebars is 18 in or 3 times the wall thickness, whichever is less.

Horizontal:

$$A_s \geq 0.03t$$

Vertical:

$$A_s \geq 0.018t$$

A_s = cross-sectional area of reinforcing bars (amount per linear foot of wall), **?** in^2/lin ft

t = thickness of wall, 9 in

Horizontal:

$$A_s \geq 0.03 \times 9 = 0.27 \text{ in}^2/\text{lin ft}$$

Use #6 at 18 in o.c.:

$$0.44 \times \frac{12}{18} = 0.29 \text{ in}^2/\text{lin ft} > 0.27 \text{ in}^2/\text{lin ft}$$

Vertical:

$$A_s \geq 0.018 \times 9 = 0.162 \text{ in}^2/\text{lin ft}$$

Use #4 at 16 in o.c.

$$0.20 \times \frac{12}{15} = 0.16 \text{ in}^2/\text{lin ft} \quad \text{Close enough}$$

NOTE: If this wall was greater than 10 in thick (unless it was a basement wall), the above amounts of reinforcing would have to be installed on *each* side of the wall at a specified distance (usually 1 to 2 in) away from each face.

Basement Wall

Example 1 A reinforced concrete basement wall in a shopping center is 10 ft 4 in tall from floor to ceiling. The top of the earth on the outside is level with the top of the wall, and the wall is nonloadbearing. How thick should this wall be, what reinforcing should it have, how should it be placed, and how much pressure will the outside earth exert upon the top of the wall? $f_c = 4000$ lb/in^2. $f_y = 60,000$ lb/in^2. (See Fig. 3-60.) The footing is adequately drained.

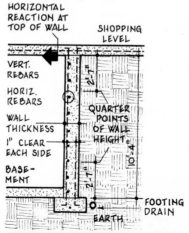

HORIZONTAL REACTION AT TOP OF WALL

SHOPPING LEVEL

VERT. REBARS

HORIZ. REBARS

WALL THICKNESS

1" CLEAR EACH SIDE

BASE-MENT

QUARTER POINTS OF WALL HEIGHT

2'-7"

10'-4"

2'-7"

FOOTING DRAIN

EARTH

FIGURE 3-60 Basement wall 1.

STEP 1. Find the wall's thickness, rebar sizes, and horizontal reaction at the top of earth (which in this case is at the top of the wall) from Table 3-24.

TABLE 3-24 PROPERTIES OF BASEMENT WALLS*

Height of wall, ft	Hor. reaction at top of earth, lb/lin ft	Hor. reaction at bottom of wall, lb/lin ft	Wall thickness, in	Vert. reinforcing, in o.c.	Horiz. reinforcing, in o.c.
6	189	378	8	#4 at 16	#4 at 16
7	257	514	8	#4 at 16	#4 at 13
8	336	671	8	#4 at 16	#4 at 11
9	425	849	8	#4 at 16	#4 at 10
10	524	1049	8	#4 at 16	#4 at 10
11	634	1269	9	#4 at 14	#5 at 13
12	755	1510	10	#4 at 13	#5 at 12
13	886	1772	11	#4 at 12	#5 at 11
14	1028	2055	12	#5 at 17	#6 at 14
15	1180	2360	13	#5 at 15	#6 at 13
16	1342	2685	14	#5 at 14	#6 at 12

*When weight of well-drained banked earth equals 110 lb/ft^3

At $H = 10$ ft. 4 in: use 11 ft

Thickness, 9 in

Vertical rebar, #4 at 14 in o.c.

Horizontal rebar, #5 at 13 in o.c.

H at top, 634 lb/lin ft

STEP 2. Arrange the horizontal and vertical rebar near the inside of the wall as shown in Fig. 3-60. Near the outside, locate the same amount of vertical rebar down from the top and up from the bottom to quarter points of the wall height. Concrete thickness from rebar to wall faces should be 1 in clear.

Example 2 A reinforced concrete basement wall of a museum has a floor-to-ceiling height of 14 ft 8 in. Earth rests against the outside to a height of 10 ft, and the total design load upon the top is 6.2 kips/lin ft. The wall is built of 3000-lb concrete and grade 60 rebars. Size the wall thickness and locate its reinforcing. (See Fig. 3-61.)

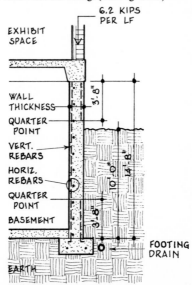

FIGURE 3-61 Basement wall 2.

STEP 1. Obtain the wall thickness needed to withstand the 10 ft tall earth pressure from Table 3-24. Use $H = 10$ ft.

Wall thickness = 8 in because of weight of earth against it.

STEP 2.   Compute the extra wall thickness to be added due to the bearing from the following formula.

$$P = 4.62 f_c T_l \left[1 - \left(\frac{H}{40t} \right)^2 \right]$$

P = total design load upon the wall, 6.2 kips/lin ft = 6200 lb/lin ft
f_c = allowable unit stress of concrete in the wall, 3000 lb/in^2
T_l = extra wall thickness due to bearing load, ? in
H = total height of wall, 14 ft 8 in = 14 × 12 + 8 = 176 in
t = thickness of wall due to earth against side, 8 in

$$6200 = 4.62 \times 3000 \times T_l \left[1 - \left(\frac{176}{40 \times 8} \right)^2 \right]$$

$$T_l = 0.64 \text{ in} \qquad \text{Use 1 in}$$

Total thickness of wall = 8 + 1 = 9 in.

STEP 3.   Compute the amount of reinforcing from the formulas below or find the amounts from Table 3-24.
Vertical rebars: C-S area/linear foot ≥ 0.018t
Horizontal rebars: C-S area/linear foot ≥ 0.03t
From Table 3-24, amount of reinforcing in a 9 in thick wall:

Vertical   #4 at 14 in o.c. 1 in from inside face, top to bottom of wall
Horizontal   #5 at 13 in o.c. 1 in from inside face, top to bottom of wall, and #5 at 13 in o.c. 1 in from outside face in top quarter and bottom quarter of wall

Retaining Walls
Retaining wall features are shown in Fig. 3-62 and in Table 3-25.

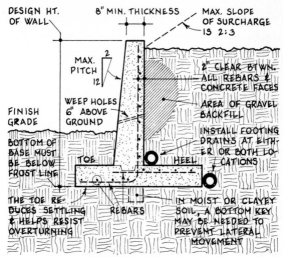

DESIGN HT. OF WALL

8" MIN. THICKNESS

MAX. SLOPE OF SURCHARGE IS 2:3

MAX. PITCH 2/12

2" CLEAR BTWN. ALL REBARS & CONCRETE FACES

AREA OF GRAVEL BACKFILL

FINISH GRADE

WEEP HOLES 6" ABOVE GROUND

INSTALL FOOTING DRAINS AT EITHER OR BOTH LOCATIONS

BOTTOM OF BASE MUST BE BELOW FROST LINE

TOE

HEEL

THE TOE REDUCES SETTLING & HELPS RESIST OVERTURNING

REBARS

IN MOIST OR CLAYEY SOIL, A BOTTOM KEY MAY BE NEEDED TO PREVENT LATERAL MOVEMENT

FIGURE 3-62 Retaining wall.

TABLE 3-25 PROPERTIES OF RETAINING WALLS

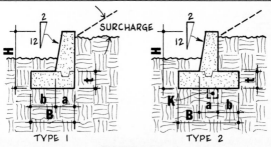

SURCHARGE

2/12

H

b   a

B

TYPE 1

2/12

H

K

c

a   b

B

TYPE 2

FIGURE 3-63

TABLE 3-25 PROPERTIES OF RETAINING WALLS *(Continued)*

H	4 to 12 ft	6 to 20 ft
B, without surcharge	$0.67H$	$0.75H$
with surcharge	$1.25H$	$0.75H$
a	$0.083H + 0.67$	$0.083H + 0.67$
b	$B - a$	$0.375H$
t	a	a
k, depth and width	—	$0.67a$
Reinforcing, minimum cross-sectional area per linear foot		
R_v	$\sqrt{0.045H}$, min. #4 at 12″ o.c. (0.20 in²/lin ft)	$\sqrt{0.04H}$, min. #4 at 12″ o.c. (0.20 in²/lin ft)
R_h	$0.36a$, min. #3 at 18″ o.c. (0.073 in²/lin ft)	$0.36a$, min. #3–18″ o.c. (0.073 in²/lin ft)
R_{lat}	$\sqrt{0.045H}$, min. #4 at 12″ o.c. (0.20 in²/lin ft)	$\sqrt{0.003H}$, min. #4 at 12″ o.c. (0.20 in²/lin ft)
R_{lon}	$0.29a$, min. #3 at 18″ o.c. (0.073 in²/lin ft)	$0.29a$, min. #3 at 18″ o.c. (0.073 in²/lin ft)

Example A retaining wall resembling type 2 in Fig. 3-63 rises 12 ft 6 in above the ground. If the local frost depth for the region is 3 ft 6 in and the ground is nearly level behind the top of the wall, what is the wall's design? $f_c = 3000$ lb/in². $f_y = 60,000$ lb/in²

Design the retaining wall from information obtained from Table 3-25.

H = height of retaining wall (from bottom of base to top of wall; H = part above ground + part extending down to below frost level), 12 ft 6 in + 3 ft 6 in = 16 ft

B = width of base (level ground behind top means no surcharge behind wall), $0.75H = 0.75 \times 16 = 12$ ft

a = width of wall at base, $0.083H + 0.67 = 0.083 \times 16.0 + 0.67$
$\quad = 1.998 = 2$ ft

b = width of heel, $0.375H = 0.375 \times 16.0 = 6$ ft

t = thickness of base, $t = a = 2$ ft

k = depth and width of bottom key (one is usually not required
$\quad$ unless there is surcharge, excessive moisture behind the wall, or
$\quad$ soil of low bearing value under the wall), $0.67a = 1$ ft 4 in

R_v = amount of vertical reinforcing in wall; minimum cross-sectional
$\quad$ area in square inches per linear foot, $R_v \geq \sqrt{0.04H} =$
$\quad \sqrt{0.04 \times 16} = 0.80$ in^2/lin ft minimum C-S area; select a
$\quad$ rebar and spacing that satisfies this requirement, maximum
$\quad$ spacing = 18 in, use #8 at 12 in o.c.

$$\text{Check:} \quad 0.79 \times \frac{12}{12} = 0.79 \qquad \text{Close enough to 0.80}$$

R_h = amount of horizontal reinforcing in wall; minimum cross-sec-
$\quad$ tional area in square inches per linear foot, $R_h \geq 0.36a = 0.36$
$\quad \times 2 = 0.72$ in^2/lin ft minimum; use #8 at 13 in o.c.:

$$\text{Check:} \quad 0.79 \times \frac{12}{13} = 0.73 \text{ in}^2/\text{lin ft} > 0.72$$

R_{lat} = amount of lateral reinforcing in base; minimum cross-sectional
$\quad$ area in square inches per linear foot, $R_{\text{lat}} \geq \sqrt{0.003H} =$
$\quad \sqrt{0.003 \times 16} = 0.219$ in^2/lin ft minimum; use #5 at 17 in o.c.

$$\text{Check:} \quad 0.31 \times \frac{12}{17} = 0.219 \text{ in}^2/\text{lin ft} \geq 0.219$$

R_{lon} = amount of longitudinal reinforcing in base; minimum cross-sec-
$\quad$ tional area in square inches per linear foot, $R_{\text{lon}} \geq 0.29a = 0.29$
$\quad \times 2 = 0.58$ in^2/lin ft minimum; use #8 at 16 in o.c.

$$\text{Check:} \quad 0.79 \times \frac{12}{16} = 0.59 \text{ in}^2/\text{lin ft} > 0.58$$

Connections

In reinforced concrete construction, great care must be exercised in assembling the meetings of beams to girders, beams or girders to columns, columns to footings, floors to tops and bottoms of stairs, and other junctions of one structural part to another. Similar care must also be exercised when the work of placing concrete in the forms stops for more than 1 hr because this delay creates a seam of weakness between the previous and subsequent batch of poured concrete. Where such seams are unavoidable, they are usually located at the junctions of structure mentioned above. Such seams must never exist in areas of positive bending moment in any structural member.

At junctions of structure, rebars are often designed to overlap so that one bar will protrude after the first part of structure is poured; then the second bar is fastened to the first before the second part is poured. At other times, small pieces or assemblies of reinforcing called *dowels* are inserted into the edge of the pour, with approximately half the dowel imbedded in the just-poured concrete and the other half protruding to become immersed in the next pour.

Dowels for reinforced concrete work have many shapes. In each, the steel's diameter, length, and spacing must be carefully designed.

Example 1: Development Length in Tension Dowels are required at the meeting of staircase top to floor girder as shown in Fig. 3-64. The rebars parallel to the dowels are #4 at 16 in o.c. in the floor slab and #5 at 8 in o.c. in the staircase. What should be the diameters and spacing of the dowels? How long should each stem of each dowel be? What is each dowel's angle of bend? $f_y = 60,000$ lb/in^2. $f_c = 3000$ lb/in^2.

Dowel diameter and spacing: Match the maximum diameter and minimum spacing of the rebars the dowels will be tied to. Use #5 rebars at 8 in o.c. All stems imbedded in the staircase will tie to a #5 bar. Every other stem imbedded in the floor slab will tie to the #4 bars at 16 in o.c., and the stems in between will lie free in the concrete.

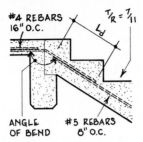

FIGURE 3-64 Rebars in top of staircase.

Length of dowel stems (development length): The dowels are subjected to tensile stress according to the following formula.

$$L_d = \frac{0.04 A f_y}{\sqrt{f_c}} \geq 8 \text{ in}$$

L_d = development length of each dowel stem for dowels subjected to tensile stress, ? in

A = cross-sectional area of the dowel rebar (from Table 3-21, C-S area of #5 rebar), 0.30 in^2

f_y = yield strength of rebar, 60,000 lb/in^2

f_c = compressive unit stress of concrete, 3000 lb/in^2

$$L_d = \frac{0.04 \times 0.30 \times 60,000}{\sqrt{3000}}$$
$$= 13.1 \quad \text{Use 14 in.}$$

Angle of bend:

Riser/tread ratio = 7/11 = tan 32.47° Use 32°.
Angle of bend = 180° − 32° = 148°

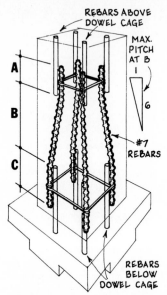

FIGURE 3-65 Rebars in column.

Example 2: Development Length in Compression A 12 x 12 in square column requires at floor level the dowel cage assembly shown in Fig. 3-65. If each vertical piece is a #7 rebar, what are the lengths of *a*, *b*, and *c*? $f_y = 60,000 \text{ lb/in}^2$. $f_c = 4000 \text{ lb/in}^2$.

Development length: The dowels are subjected to compressive stress; minimum lengths of *a*, *b*, and *c* = 8 in or $0.003 f_y d$, whichever is greater; *a* = *c*.

$$L_d = \frac{0.02 d f_y}{\sqrt{f_c}}$$

L_d = development length of stems a and c for dowels subjected to compressive stress, ? in

d = diameter of dowel (from Table 3-21, diameter of #7 rebar), 0.875 in

f_y = yield strength of steel, 60,000 lb/in^2

f_c = compressive unit stress of concrete, 3000 lb/in^2

$$L_d = \frac{0.02 \times 0.875 \times 60,000}{\sqrt{3000}}$$
$$= 19.2 \text{ in}$$

Use 20 in = length of a and c.

   Check:

$$L_d \geq 0.0003 f_y d$$
$$20 \geq 0.003 \times 60,000 \times 0.875$$
$$20 \geq 15.75 \text{ in} \quad \text{OK}$$

Length of b:

$$b = 3dp$$

d = diameter of dowel, 0.875 in

p = pitch of segment b; assume minimum pitch of 6:1:

$$b = 3 \times 0.875(6/1) = 6.89 \text{ in}$$

Use 7 in.

Example 3: Hook Design   The tensile reinforcing in a 12 x 32 in concrete beam has hooks at each end as shown in Fig. 3-66. If the reinforcing is #8 rebar, what are the dimensions of the hooks?

   Inside diameter:

#3 to #8 bars: $i = 6d$
#9 to #11 bars: $i = 8d$

FIGURE 3-66   Rebar hook.

i = inside diameter of hook, ? in
d = diameter of rebar (from Table 3-21, diameter of #8 rebar), 1 in

$$i = 6 \times 1.00 = 6 \text{ in}$$

Extension length:

$$e = 4d \geq 2.5 \text{ in}$$

e = length of extension, ? in
d = diameter of rebar, 1.00 in

$$e = 4 \times 1 = 4 \text{ in}$$

FOOTINGS

Footing design begins with familiarity with the ground that will lie under the footings. (Properties of various types of soils are given in Table 3-26.) The following ground surface characteristics may help in ascertaining the nature of earth several feet underneath.

Nearby existing buildings   If the new footings are deep and near existing buildings, possible severe lateral earth movement may occur. Shoring of earth or temporary support of existing foundations may be required.

TABLE 3-26 PROPERTIES OF SOILS

Type of soil	Symbol, (Fig. 3-67)	Bearing, kips/ft² *	Drainage ability
Solid hard granite or gniess		80–160	None
Solid limestone, slate		50–80	None
Soft limestone		24–30	None
Hardpan		16–20	Poor
Boulders with rocks or sand		12–16	Good
Rotten or loose rock		10	Fair
Compact uniform gravel		10	Excellent
Firm dry clay		8	Poor
Clean compact sand		7	Excellent
Rocks with clay		6	Poor
Rocks with organic soil		6	Good
Clay with sand or silt		4	Fair
Soft wet clay		3	Poor
Clay with organic soil or silt		2	Fair
Peat, topsoil, organic soil		2	Good
Mud, quicksand, flowing soils		0–0.5	Poor

*When performing calculations involving soil bearing capacities, obtain test results for all critical designs.

Rock outcrops Possible bedrock may exist just under the surface. This is good for bearing and frost resistance but bad for excavating.

Foliage Some species (willow, skunk cabbage) indicate very moist soil. Low, dense vegetation may mean moist, porous, weak soil. Absence or thinness of foliage in a verdant area suggests hard, firm soil. Large, solitary trees indicate solid ground.

Bodies of water Nearby surrounding land most likely has high water table close to ground surface. Footing construction probably will be expensive and the final result unstable.

Level topography Probably easy excavating, lower than average bearing, poor drainage.

Gentle slope Probably easy excavating and good drainage.

Convex terrain Topography forms slight ridge in sloping terrain. Ground and surface water drains away from these areas. Usually a dry, solid place to build.

Concave terrain Topography forms slight valley in sloping terrain. Water drains into these areas. Almost always a wet, soft, weak place to build.

Steep slopes Costly excavating, erosive damage, sliding soils.

Soil Bearing Capacity

Soil borings indicate that a 3-ft thick layer of hardpan exists over a stratum of soft wet clay under where a square footing must be built. If the total load on the footing is 100 kips, including the footing's weight, what is the footing's area?

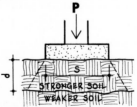

FIGURE 3-68 Footing cross section 1.

STRATEGY Although the footing's load will eventually rest on the weaker soil, some of the weight will spread outward through the depth of the stronger soil, as shown in Fig. 3-68. This condition is represented by the two formulas below. The maximum value governs.

$$S = \sqrt{\frac{P}{C_w}} - 2d \sqrt{\frac{C_s}{160}} \qquad S \geq \sqrt{\frac{P}{C_s}}$$

S = length of one side of the square footing, ? ft
P = total actual load on the footing, including the footing's weight, 100 kips
C_w = bearing capacity for the weaker underlying soil (from Table 3-26, C for soft wet clay), 3 kips/ft^2
d = depth of stronger upper soil, 3 ft
C_s = bearing capacity of the stronger upper soil (from Table 3-24, C for hardpan), minimum 16 kips/ft^2

$$S \geq \sqrt{\frac{100}{3}} - 2 \times 3 \sqrt{\frac{16}{160}} = 3.88 \text{ ft minimum} +$$

$$S \geq \sqrt{\frac{100}{16}} = 2.5 \text{ ft minimum}$$

The larger value (3.88) governs. Use 4.00 ft.
   Area of footing:

$$A = S^2 = 4.00^2 = 16.0 \text{ ft}^2$$

NOTE: When a ground surface is covered with a thin layer of material having a stronger bearing value, in any calculations use the bearing value of the weaker, underlying soil. For example, 4 in of compact uniform gravel laid on soft wet clay has a bearing capacity not of 10 kips/ft^2 but 3 kips/ft^2.

Continuous Wall Footing

The live load on a 12-in thick wall is 6.2 kips/lin ft, and the dead load is 9.6 kips/lin ft. If the earth under the wall has a bearing capacity of 4000 lb/ft^2, what is the size of the wall's footing? How much reinforcing should the footing have? f_c = 3000 lb/in^2. f_y = 60,000 lb/in^2. (See Fig. 3-69.)

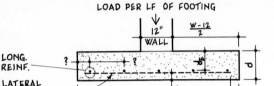

FIGURE 3-69 Footing cross section 2.

STEP 1. Determine the width of the footing.

$$W = \frac{12P}{B}$$

W = width of the footing, **?** in

P = total load on the footing per linear foot; equals live load + dead load + estimated weight of the footing; $P \approx 1.07(\text{DL} + \text{LL})$: DL = 9.6 kips/lin ft, LL = 6.2 kips/lin ft; $1.07(9.6 + 6.2) = 16.9$ kips/lin ft = 16,900 lb/lin ft.

B = bearing capacity of the soil, 4000 lb/ft^2

$$W = \frac{12 \times 16,900}{4000} = 50.7 \text{ in}$$

Use 52 in.

STEP 2. Calculate the depth of the footing.

$$2\sqrt{f_c} = \frac{P_d(W - b - 2d_e)}{20.4Wd_e}$$

f_c = allowable concrete unit stress, 3000 lb/in^2

P_d = design load on the wall, per linear foot:

$P_d = 1.4D + 1.7L \qquad D = 9.6$ kips/lin ft, $L = 6.2$ kips/lin ft
$\quad = 1.4 \times 9.6 + 1.7 \times 6.2 = 24$ kips/lin ft
$\quad = 24,000$ lb/lin ft

W = width of footing, 52 in
b = width of wall, 12 in
d_e = effective depth of footing (equals the distance from the top of the footing to the center of the lateral reinforcing), ? in

$$2\sqrt{3000} = \frac{24{,}000(52 - 12 - 2d_e)}{20.4 \times 52d_e}$$
$$116{,}000d_e = 960{,}000 - 48{,}000d_e$$
$$d_e = 5.85 \text{ in}$$

Total depth:

$$d = d_e + 3.5$$
$$= 5.85 = 3.5 = 9.35 \text{ in} \qquad \text{Use 10 in.}$$

Actual $d_e = 10 - 3.5 = 6.5$ in.

STEP 3. Calculate the footing's lateral reinforcing (the rebars running perpendicular to the wall) from the two formulas below. The larger value governs. Maximum spacing of rebars is 18 in o.c.

$$A_s \geq \frac{P_d(W - b)^2}{6.48\,Wd_ef_y} \qquad A_s \geq \frac{2400d_e}{f_y}$$

A_s = amount of lateral reinforcing (cross-sectional area per linear foot), ? in^2/lin ft
P_d = 24,000 lb
W = 52 in
b = 12 in
d_e = 6.5 in
f_y = 60,000 lb/in^2

$$A_s \geq \frac{24{,}000(52 - 12)^2}{6.48 \times 52 \times 6.5 \times 6{,}000} = 0.29 \text{ in}^2/\text{lin ft} \quad +$$

$$A_s \geq \frac{2400 \times 6.5}{60{,}000} = 0.26 \text{ in}^2/\text{lin ft}$$

Use #6 at 18 in o.c.:

$$0.44 \times 12/18 = 0.293 > 0.29$$

STEP 4. Calculate the footing's longitudinal reinforcing.

$$A_s \geq 0.002Wd$$

A_s = amount of longitudinal reinforcing (cross-sectional area for the total width of the footing), ? $\text{in}^2/\text{lin ft}$

W = 52 in

$$d = 10 \text{ in}$$
$$A_s \geq 0.002 \times 52 \times 10 = 1.04 \text{ in}^2$$

Use six #4 at 11 in o.c.:

$$0.20 \times 6 = 1.20 > 1.04$$

Column Footing

A 10 x 10 in column supports a live load of 72 kips and a dead load of 88 kips; the soil it rests on is clean, compact sand. If the footing under this column is square, what is the length of a side, the length of its depth, and its arrangement of reinforcing? $f_c = 3000 \text{ lb/in}^2$. $f_y = 60,000 \text{ lb/in}^2$.

STEP 1. Compute the length of the footing's side.

$$S = \sqrt{\frac{P}{C}}$$

S = length of one side of the square footing, ? in

P = total service load on the footing; equals live load + dead load + estimated weight of footing; $P \approx 1.07 \text{ (DL + LL)}$: DL = 88 kips, LL = 72 kips; $P = 1.07(88 + 72) = 171$ kips

C = bearing capacity of the soil under the footing (from Table 3-22, bearing capacity of clean, compact sand is 7 kips/ft^2),

$$S = \sqrt{\frac{171}{7}} = 4.94 \text{ ft}$$

Use 5 ft = 60 in.

STEP 2. Select a trial depth. Try $d = 14$ in

STEP 3. Check the depth for punching shear. This occurs at the dotted line in Fig. 3-70

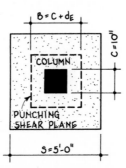

FIGURE 3-70 Column footing punching shear.

$$\text{Actual } v_h = \frac{P_d(S^2 - B^2)}{3.4S^2Bd_e}$$

$$\text{Allowable } v_h = 4\sqrt{f_c}$$

v_h = unit concrete shear, ? lb/ft^2

P_d = total design load on the footing, 1.4DL + 1.7LL

    DL = dead load on footing = 88 kips = 88,000 lb

    LL = live load on footing = 72 kips = 72,000 lb

    P_d = 1.4 × 88,000 + 1.7 × 72,000 = 246,000 lb

S = length of one side of the footing from step 1, 60 in

d = depth of beam, $d = 14$ in

B = width of punching shear plane, width of column + d = 10 + 14 = 24 in

f_c = allowable concrete unit stress, 3000 lb/in²

$$\text{Actual } v_h = \frac{246,000(60^2 - 24^2)}{3.4 \times 60^2 \times 24 \times 12.5} = 203 \text{ lb/in}^2$$

$$\text{Allowable } v_h = 4\sqrt{3000} = 219 \text{ lb/in}^2 \quad \text{OK}$$

STEP 4. Check the depth for beam shear. This occurs at the dotted line in Fig. 3-71.

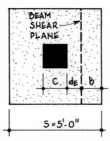

FIGURE 3-71 Column footing beam shear.

$$\text{Actual } v_h = \frac{bP_d}{S^2 d}$$

$$\text{Allowable } v_h = 2\sqrt{f_c}$$

v_h = unit shear for the concrete in the footing, ? lb/in²

b = width of beam shear plane, $0.5(S - C - d_e)$ = 0.5(60 − 10 − 2 × 14) = 11 in

P_d = 246,000 lb

S_c = 60 in

d_e = 14 in

f_c = 3000 lb/in²

$$\text{Actual } v_h = \frac{11 \times 246{,}000}{60^2 \times 14} = 53.7 \text{ lb/in}^2$$

$$\text{Allowable } v_h = 2\sqrt{3000} = 110 \text{ lb/in}^2 \qquad \text{OK}$$

STEP 5. Design the reinforcing. As the footing is square, the rebars are the same both ways. There are two criteria, represented by the formulas below. The larger number governs. Maximum spacing of rebars is 18 in o.c. (See Fig. 3-72.)

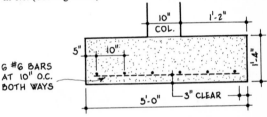

FIGURE 3-72 Footing cross section 3.

$$A_s \geq \frac{0.138 P_d (S - C)^2}{S f_y (d_e - 1)} \qquad A_s \geq \frac{200 S d_e}{f_y}$$

A_s = minimum cross-sectional area of reinforcing each way in the footing, ? in^2

P_d = 246,000 lb

S = 60 in

C = 10 in

f_y = 60,000 lb/in^2

d_e = effective depth of beam $d - 3.5 = 14 - 3.5 = 10.5$ in

$$A_s \geq \frac{0.138 \times 246{,}000 (60 - 10)^2}{60 \times 60{,}000 (10.5 - 1)} = 2.25 \text{ in}^2$$

$$A_s \geq \frac{200 \times 60 \times 13}{60{,}000} = 2.60 \text{ in}^2$$

Use six #6 at 10 in o.c. each way:

$$6 \times 0.44 = 2.64 \text{ in}^2 > 2.60 \text{ in}^2$$

Combined Footings

Two 21-in round columns are located as shown in Fig. 3-73. Column A supports 195 kips and column B supports 127 kips. If the soil underneath is clayey silty soil, what is the length, width, and depth of the combined footing under the two columns? How should the reinforcing be arranged?

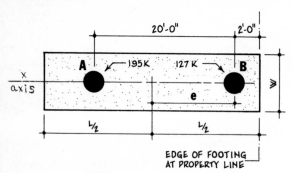

FIGURE 3-73   Combined footing plan.

STEP 1.   Locate the centroid of the column loads by taking moments about B.

$$P_a S = e(P_a + P_b)$$

P_a = point load on column A, 195 kips
P_b = point load on column B, 127 kips
S = distance between centers of columns A and B, 20 ft
e = distance between column load centroid and center of column B, ? ft

$$195 \times 20 = e(195 + 127)$$
$$e = 12.1 \text{ ft}$$

STEP 2. Determine the length of the footing.

$$L = 2(e + s)$$

L = total length of footing under both columns, ? ft

e = distance between column centroid and center of column B from step 1, 12.1 ft

s = distance between center of column B and property line, 2 ft

$$L = 2(12.1 + 2) = 28.2 \text{ ft}$$

Use 28 ft 4 in (28.3 ft).

STEP 3. Calculate the width of the footing.

$$P = WLC$$

P = total actual load on the footing (equals column loads + weight of footing; weight of footing $\approx 1.15 \times$ column loads), 1.15(195 + 127) = 370 kips

W = width of footing, ? ft

L = length of footing from step 2, 28.3 ft

C = bearing capacity of soil (from Table 3-26, bearing capacity of clayey silty soil), 4 kips/ft^2

$$370 = W \times 28.3 \times 4$$
$$W = 3.27 \text{ ft}$$

Use 3 ft 4 in (3.33 ft).

STEP 4. Estimate the depth of the footing.

$$D \approx 0.09S$$

D = estimated depth of footing (this is a preliminary assumption), ? ft

S = distance between centers of columns, 20 ft

$$D \approx 0.09 \times 20 = 1.80 \text{ ft}$$

Use 1 ft 10 in

STEP 5. Arrange the reinforcing as shown in Fig. 3-74.

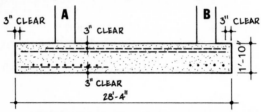

FIGURE 3-74 Combined footing cross section.

NOTE: Further analysis is quite complicated. Basically, it involves:

1. Drawing shear and moment diagrams through the X axis of the footing
2. Calculating maximum moment, one-way shear, and two-way shear at critical sections
3. Checking the estimated depth for one-way shear, two-way shear, and maximum moment.
4. Designing the top longitudinal reinforcing
5. Designing the bottom longitudinal reinforcing
6. Designing the bottom lateral reinforcing

Piles: Friction Type

A tapered fluted pile, made of a tapered hollow metal shaft filled with concrete, is driven until the pile driver, a 1540-lb hammer falling a distance of 12 ft, has given the shaft an average of 1.3 in penetration for each of the last five blows. What is the bearing capacity of the pile?

$$P = \frac{2WH}{S + 1}$$

P = bearing capacity of the pile, ? lb
W = weight of the pile driver hammer, 1540 lb
H = height of drop of hammer, 12 ft
S = pile penetration under final blow, or average of 5 final blows, 1.3 in

$$P = \frac{2 \times 1540 \times 12}{1.3 + 1} = 16,100 \text{ lb}$$

Say 16 kips.

Piles: Bearing Type

Soil test borings indicate that 48 ft below the foundation level of a new building exists granite bedrock. Piles of open-end steel pipe filled with concrete will pass through the uppermost 48 ft of soft soil to support the building. If the steel pipe is 12 in in diameter, what is the bearing of each pile if the pipe is driven to refusal?

$$P = 0.785d^2C$$

P = bearing capacity of the pile, ? kips
d = diameter of the pile, 12 in = 1 ft
C = bearing capacity of the soil of refusal (from Table 3-26, bearing capacity of solid hard granite is 80–160 kips/ft^2; to be safe, use 80 kips/ft^2

$$P = 0.785 \times 1.0 \times 80$$
$$= 62.8 \text{ kips}$$

Piles: Belled-Caisson Type

Soil test borings indicate that a stratum of hard limestone exists 65 ft below the level of pile-cap footings for a new office building. Considering that the local building code limits caisson shaft heights to 30 times the shaft diameter, what are the dimensions of the smallest belled-caisson pile that can be built? What weight will it support?

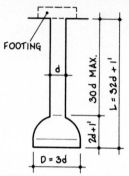

FIGURE 3-75 Caisson cross section.

STEP 1. Find the diameter of the caisson. (See Fig. 3-75.)

$$L = 32d + 1$$

L = total length of caisson shaft and bell (equals distance from ground surface to limestone), 65 ft
d = diameter of shaft, ? ft

$$65 = 32d + 1$$
$$d = 2 \text{ ft}$$

STEP 2. Calculate the load the caisson will support.

$$P = 0.785CD^2$$

P = total load supported by the caisson, ? kips
C = bearing capacity of the soil the caisson bell rests on (from Table 3-22, C for solid limestone), 50–80 kips/ft²; use 50 kips/ft²
D = diameter of caisson bell, $D = 3d = 3 \times 2 = 6$ ft

$$P = 0.785 \times 50 \times 6^2$$
$$= 1410 \text{ kips}$$

Pile-Cap Footings

A 16-in square column supports a dead load of 32 kips and a live load of 19 kips. If this load is transferred to piles having a capacity of 16 kips each, what is the design of the footing between the column and the piles? What reinforcing should the footing have? $f_c = 3000$ lb/in^2. $f_y = 60,000$ lb/in^2.

STEP 1. Determine the number of piles.

$$N = \frac{1.15(DL + LL)}{C}$$

N = number of piles, ? piles
DL = dead load on the column, 32 kips
LL = live load on the column, 19 kips
C = bearing capacity of the piles, 16 kips each
  (*Note:* The 1.15 in the formula is a 15 percent footing weight factor.)

$$N = \frac{1.15(32 + 19)}{16}$$

$$= 3.67$$

Use four piles.

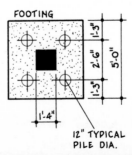

FIGURE 3-76 Footing showing bearing piles.

STEP 2.   Design the footing in plan according to the following criteria (see Fig. 3-76):

a.   The centers of the piles must be at least 2 ft 6 in apart.
b.   The distance from the centers of the piles to the edge of the footing must be at least 1 ft 3 in.

NOTE: Diagrams of pile-cap footings for arrangements of from one to seven piles are given in Fig. 3-79.

STEP 3.   Determine the design load on each pile.

$$P_d = \frac{1.4DL + 1.7LL}{N}$$

P_d = design load on each pile, ? kips
DL = 32 kips
LL = 19 kips
N = 4 piles

$$P_d = \frac{1.4 \times 32 + 1.7 \times 19}{4}$$

$$= 19.3 \text{ kips}$$

STEP 4.   Calculate the effective depth of the footing. Minimum d_e = 12 in. (Refer to Fig. 3-77.)

FIGURE 3-77   Footing plan showing piles and beam shear.

$$2\sqrt{f_c} = \frac{V}{0.85Sd_e}$$

f_c = allowable concrete unit shear, 36,000 lb/in²

V = total shear between column and piles (maximum shear occurs at dotted line in Fig. 3-77; at shear plane, two piles resist up as column acts down), $V = 2 \times 16$ kips = 32 kips = 32,000 lb

S = width of shear plane in inches (equals length of side, from diagram of four-pile footing in Fig. 3-79), 5 ft = 60 in

d_e = depth of shear plane, ? in (must be 12 in minimum)

$$2\sqrt{3000} = \frac{32,000}{0.85 \times 60 d_e}$$

$$d_e = 5.73 \text{ in} < 12 \text{ in} \qquad \text{Use 12 in.}$$

STEP 5. Calculate the amount of reinforcing. As the footing is square, the cross-sectional area of reinforcing is the same both ways. Maximum spacing of rebars equals 18 in o.c. (See Fig. 3-78.)

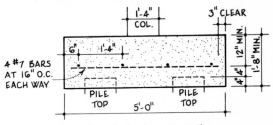

FIGURE 3-78 Footing cross section.

$$A_s \geq \frac{200 S d_e}{f_y}$$

A_s = minimum cross-sectional area of reinforcing each way in the footing, ? in²

S = 60 in

d_e = 12 in

f_y = 60,000 lb/in²

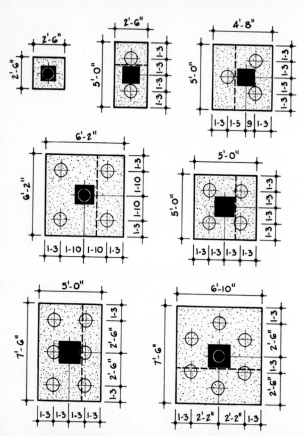

FIGURE 3-79 Pile-top footing arrangements.

$$A_s \geq \frac{200 \times 60 \times 12}{60,000} = 2.40 \text{ in}^2$$

Use four #7 at 16 in o.c.:

$$\text{Check:} \quad 4 \times 0.60 = 2.40 \text{ in}^2 \geq 2.40 \text{ in}^2$$

Pile-top footing arrangements are illustrated in Fig. 3-79. The dotted lines indicate the location of maximum shear planes.

MASONRY

Masonry includes brick, concrete block, field stones, cut stones, ceramic tile, and glass block. These materials are usually laid in an interlocking fashion to produce a solid mass. The strength of such construction depends on the strength of the units employed and the strength of the mortar used to hold the units together. Usually the minimum value governs.

Loads on a masonry wall should bear on the center tenth of the wall's thickness (eccentricity = 0.05); otherwise, the wall's strength is considerably reduced. For example, if the load center is on the edge of the center fifth of the width ($E = 0.10$), the wall is about 15 percent weaker. (Refer to Fig. 3-80.)

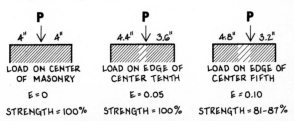

FIGURE 3-80 Three masonry loads.

TABLE 3-27 ALLOWABLE MASONRY STRESSES*

Type of stress and masonry unit or condition	Type of mortar		
	N	S	M
	Allowable stress, lb/in²		
Compression, f_c, lb/in²			
Brick, SW	300	350	400
Brick, MW	275	310	350
Brick, NW	215	235	290
Concrete block, grade A walls	85	90	100
Concrete block, grade B walls	70	75	85
Concrete block, grouted piers	90	95	105
Cut granite	640	720	800
Cut limestone, marble	400	450	500
Cut sandstone, cast stone	320	360	400
Rubble, rough, random	100	120	140
Glass block, min. 3 in thick			
Exterior walls: Unsupported surface area ≤ 144 ft²			
Unsupported length ≤ 25 ft			
Unsupported height ≤ 20 ft			
Interior walls: Unsupported surface area ≤ 250 ft²			
Unsupported length and unsupported height ≤ 25 ft			
Tension, f_t			
Normal to bed joints	28	36	36
Parallel to bed joints	56	72	72
Shear, v			
Solid units	56	80	80
Hollow units, total C-S area	19	27	27
Modulus of elasticity, $E = 800 f_c$ but not $> 3,000,000$			

*For inspected work

Bearing Wall Strength

A 10-ft 4-in tall masonry wall is made of 8-in grade A concrete block and type N mortar. How much weight per linear foot will the wall support? How much weight will the wall support if it is made of 10-in block? (Refer to Table 3-27 for a listing of allowable masonry stresses.)

$$W = CSA$$

W = total weight per linear foot the wall will support, ? lb/lin ft

C = bearing capacity of masonry (as listed in Table 3-27, C for concrete block, grade A walls, type N mortar), 85 lb/in^2

S = slenderness ratio:

$$S = 1.2 - \frac{H}{37t} \leq 1.0$$

$$H = 10 \text{ ft } 4 \text{ in} = 124 \text{ in}$$

$$t = 7.5 \text{ in}$$

$$S = 1.2 - \frac{124}{37 \times 7.5} = 0.75$$

A = cross-sectional area per linear foot of wall (8-in concrete block is 7½ in thick), 7.5 × 12 in/lin ft = 90 in^2/lin ft

$$W = 85 \times 0.75 \times 90 = 5740 \text{ lb/lin ft}$$

10-in wall: 10-in concrete block is 9.5 in thick.

$$S = 1.2 - \frac{124}{37 \times 9.5} = 0.85$$

$$A = 9.5 \times 12 = 114 \text{ in}^2/\text{lin ft}$$

$$W = 85 \times 0.85 \times 114$$

$$= 8240 \text{ lb/lin ft}$$

Bearing Wall Thickness

A brick bearing wall has an unsupported height of 12 ft. What is the wall's minimum allowable thickness?

Solid masonry:

$$t \geq 0.05H +$$

Hollow masonry:

$$t \geq 0.0556H$$

t = minimum thickness of masonry wall, ? in
H = height of masonry wall, 12 ft = 144 in

$$t \geq 0.05 \times 144 = 7.2 \text{ in}$$

Reinforcing

A 10-in thick reinforced grouted brick wall has rebars both horizontally and vertically in the grouted space between the two wythes. What is the minimum required reinforcing in each direction?

Minimum horizontal or vertical reinforcing:

$$A_s \geq 0.0084t/\text{lin ft each way}$$

Minimum total reinforcing:

$$A_s \geq 0.024t/\text{lin ft total both ways}$$

A_s = cross-sectional area of reinforcing, ? in^2/lin ft
t = thickness of wall, 10 in

For either direction, compute the horizontal first.

$$A_s = 0.0084 \times 10 = 0.084 \text{ in}^2/\text{lin ft}$$

Use #4 at 28 in o.c.:

$$0.20 \times 12/28 = 0.085 \text{ in}^2 > 0.084 \text{ in}^2$$

Total reinforcing: Vertical = total − horizontal

$$A_s = 0.024 \times 10 - 0.085 = 0.155 \text{ in}^2/\text{lin ft}$$

Use #5 at 24 in o.c.:

$$0.31 \times 12/24 = 0.155 = 0.155$$

NOTE: Maximum spacing of rebars in masonry walls is 6 times the thickness or 48 in, whichever is less.

Anchor Bolt Shear

An 8-in thick concrete block basement wall has a horizontal reaction of 189 lb/lin ft along its top as a result of the earth resting against its outer face. Anchor bolts spaced 4 ft on centers hold the sill plate of the wood framing above to the top of the wall. If the top two courses of the concrete block are grouted, how thick should the anchor bolts be? How deep should their shanks be buried in the grout?

$$HL = S$$

H = horizontal reaction at the top of the wall, 189 lb/lin ft
L = length of wall per anchor bolt, = 4 ft
S = allowable shear for each anchor bolt (from Table 3-28), **?** lb

$$189 \times 4 = 756 \text{ lb per bolt}$$

From Table 3-28, a ⅝-in bolt anchored in grouted masonry will resist a lateral force of 750 lb. This is close enough to 756 lb to be OK.

Length of imbedment: From Table 3-28, the bolt should be buried at least 4 in in the wall.

TABLE 3-28   ALLOWABLE SHEAR FOR ANCHOR BOLTS

Bolt diameter, in	Minimum imbedment, in	Solid masonry, lb	Grouted masonry or reinforced concrete, lb
½	4	350	550
⅝	4	500	750
¾	5	750	1100
⅞	6	1000	1500
1	7	1250	1850
1⅛	8	1500	2250

PROJECT DESIGN WEATHER DATA

DATA	AMOUNT
LOCAL LATITUDE	_____ °
MAXIMUM RAINFALL INTENSITY	_____ in/hr
MAXIMUM WIND PRESSURE	_____ lb/ft^2
EARTHQUAKE ZONE	_____
AVERAGE ANNUAL TEMPERATURE	_____ °F
WINTER DESIGN TEMPERATURE	_____ °F
SUMMER DESIGN TEMPERATURE	_____ °F
SUMMER DESIGN RELATIVE HUMIDITY	_____ %
SUMMER DESIGN ENTHALPY	_____ BTU/lb
LOCAL CLARITY FRACTION	_____
SUN ANGLE DUE SOUTH NOON JUN 21	_____ ° alt.
SUN ANGLE DUE SOUTH NOON DEC 21	_____ ° alt.

INSOLATION DATA, BTU/ft^2 per day

Nov 21	Dec 21	Jan 21	Feb 21	Mar 21
_____	_____	_____	_____	_____

Climate data checklist. This data box may be duplicated and attached to drawings or specifications.

CLIMATE

GENERAL

Sunshine, temperature, rainfall, wind, earthquakes, and other climatic forces shape architecture in many ways. The mathematics involved are often in two parts. The first part quantifies the nature of the climatic force, and the second sizes the architecture that resists the force.

SOLAR ANGLES

Solar angles depend upon local latitude, month of year, and time of day. Table 4-1 describes the relationship of sun angles to local latitudes and the 21st day of each month.

Altitude

On a property located at N41°23′ latitude, what is the maximum vertical angle the sun makes with a level part of the land on February 21?

TABLE 4-1 SOLAR ALTITUDE FORMULAS*

Date	Altitude in degrees, noon due south
December 21	$\angle = \ \ 66.5$—local latitude
January or November 21	$\angle = \ \ 69.5$—local latitude
February or October 21	$\angle = \ \ 78.5$—local latitude
March or September 21	$\angle = \ \ 90.0$—local latitude
April or August 21	$\angle = 101.5$—local latitude
May or July 21	$\angle = 110.5$—local latitude
June 21	$\angle = 113.5$—local latitude

*Angle of the sun above the horizon at noon due south on the 21st day of each month.

From Table 4-1, the correct formula is

$$\angle = 78.5 - L$$

$\angle$ = altitude of the sun at noon due south on February 21, **?** °
L = local latitude, 41°23′, use 41.5°

$$\angle = 78.5 - 41.5$$
$$= 37°$$

Altitude and Azimuth

What is the altitude and azimuth (compass direction angle) of the sun at 4 p.m. on May 21 in Savannah, Georgia?

STEP 1.   Find Savannah's latitude on a map. It is approximately 32°.

STEP 2.   Find the sun's altitude and azimuth for 4 p.m. May 21 on the sun disk for 32° latitude in Fig. 4-1. Locate the intersection of date and time on the heavy black lines, then read the altitude and azimuth on the light circle and radius lines. The answer is 35° above the horizon and 93° west of south.

NOTE: When using the sun disks, interpolate for intermediate values.

Angle of Incidence

A roofed surface is slanted at 60° and faces south-southeast. What angle does the sun make with the roof when the sun is 50° above the horizon and its bearing angle is 45° west of south? (Refer to Fig. 4-2.)

$$\sin A = \sin H \cos S + \cos H \sin S \cos (D - F)$$

A = angle between incident sunrays and collector surface, **?** °
H = height of sun above horizon, 50°
S = slant of roof to horizontal, 60°
D = direction of sun from due south, 45°
F = direction from due south the roof faces, 22.5°

$$\sin A = \sin 50° \cos 60° + \cos 50° \sin 60° \cos (45 - 22.5)°$$
$$= 0.897$$
$$A = 63.8°$$

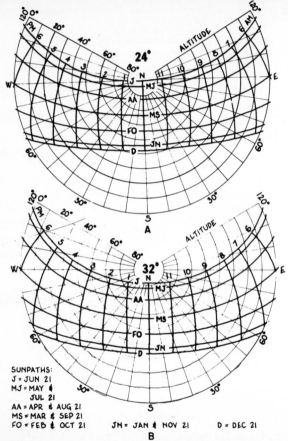

FIGURE 4-1 (*a*) 24° sun disk. (*b*) 32° sun disk. (*c*) 40° and 48° sun disks. (*From Ramsey and Sleeper, Architectural Graphic Standards, 1981.*)

CONCENTRIC CIRCLES = SUN'S ALTITUDE
RADII = SUN'S AZIMUTH
HORIZONTAL DARK CURVED LINES = MONTHLY SUNPATHS
VERTICAL DARK CURVED LINES = DAYLIGHT HOURS

C

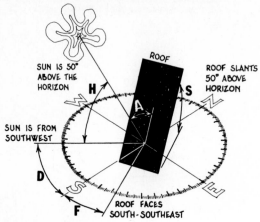

SUN IS 50°
ABOVE THE
HORIZON

ROOF SLANTS
50° ABOVE
HORIZON

SUN IS FROM
SOUTHWEST

ROOF FACES
SOUTH-SOUTHEAST

FIGURE 4-2 Sun angle on roof.

Overhang Design

Example 1 A roof extends 48 in over a vertical facade oriented due south. If the sun's angle above the horizon at noon is 35°, how much of the facade is cast in shade? (See Fig. 4-3.)

$$V = H \tan S$$

V = vertical dimension (the length of the shadow cast upon the facade, measured down from the overhang), **?** in

H = horizontal dimension (the length of the overhang), 48°

S = angle the sun's rays make with the horizontal, 35°

$$V = 48 \tan 35° = 48 \times 0.70 = 33.6 \text{ in}$$

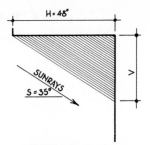

FIGURE 4-3 Overhang.

Example 2 A roof extends 30 in over a vertical facade oriented south-southeast. When the sun is from the south-southwest, its altitude is 52 degrees. At this time, how much of the facade is cast in shade?

$$H \tan S = V \cos D$$

H = horizontal dimension (the length of the overhang), 30 in
S = angle of sun's rays above the horizon, 52°
V = vertical dimension (the length of the shadow upon the facade), ? in
D = difference between the direction of the sun and the direction the facade faces. The sun is from the south-southwest, which is 22.5° west of south, whereas the facade faces south-southeast, which is 22.5° east of south. $D = 22.5 + 22.5 = 45°$

$$30 \tan 52° = V \cos 45°$$
$$V = 54.3 \text{ in}$$

TEMPERATURE

Owing to temperature changes, no material in architecture is ever still. Table 4-2 lists the coefficients of thermal expansion for various building materials.

TABLE 4-2 EXPANSION COEFFICIENTS FOR BUILDING MATERIALS

Material	Rate of expansion, "/" per °F	
Masonry		
Brick		3.5×10^{-6}
Cement, portland		7.0
Concrete		6.5
Glass		4.7
Granite, gneiss, slate		4.0
Limestone		3.8
Marble		5.6
Plaster		9.2
Sandstone		4.4
Tile, structural clay		3.3
Metals		
Aluminum, wrought		12.8×10^{-6}
Copper		9.8
Iron, cast		5.9
Iron, wrought		6.7
Lead		15.9
Steel, structural		6.5
Zinc, rolled		17.3
Plastics		
Acrylics		40 to 50×10^{-6}
Fiberglass		10 to 14
Phenolics		8.5 to 25
Woods	*Parallel to grain*	*Normal to grain*
Fir	2.1×10^{-6}	32×10^{-6}
Maple	3.6	27
Oak	2.7	30
Pine	3.0	19
Plywood	3.4	3.4

Fahrenheit to Celsius

20° Celsius (C) is how many degrees Fahrenheit (F)?

$$F = 9/5 C + 32$$

F = number of degrees on Fahrenheit scale, $?°F$
C = number of degrees on Celsius scale, $?°C$

$$F = 9/5 \times 20 + 32$$
$$= 68°F$$

NOTE: The absolute temperature scale for degrees Fahrenheit is known as Rankine: $0°F = 460°R$. The absolute temperature scale for degrees Celsius is known as Kelvin: $0°C = 273$ K.

Radiant Energy

How much more radiant energy is given off by a cast-iron stove burning oak, whose surface temperature measures 800°F., than the same stove burning pine, whose surface temperature measures 625°F?

$$R = \left(\frac{460 + F_1}{460 + F_2}\right)^4$$

R = ratio of radiant energy given off by the hotter condition to the amount given off by the cooler condition, $?$
F_1 = temperature of the hotter condition, $800°F$
F_2 = temperature of the cooler condition, $625°F$

$$R = \left(\frac{460 + 800}{460 + 625}\right)^4$$
$$= 1.82 \text{ times more radiance for the stove burning oak}$$

Thermal Expansion

Example 1 A 40-ft length of copper pipe warms from 40° to 180°F when hot water flows through it. When this happens, how much does this section of pipe lengthen?

$$I = LKT$$

I = amount of increase in the pipe's length, ? in
L = length of pipe at 40°F, 40 ft = 480 in
K = coefficient of thermal expansion for copper (from Table 4-2, K for copper = 9.8×10^{-6}), 0.0000098 in/in·°F
T = temperature change in the pipe (40° to 180°F), 180 − 40 = 140°F

$$I = 480 \times 0.0000098 \times 140$$
$$= 0.659 \text{ in}$$

Example 2 A 320-ft long brick wall is subjected to a possible maximum temperature of 180°F while exposed to direct sunlight in summer and a possible minimum temperature of −30°F on a cold winter night. If the maximum width of its expansion joints is 1 in, how many expansion joints should the wall have?

$$I = LKT$$

I = length of increase in the expansion joint from lowest to highest temperature, 1 in
L = length of wall per expansion joint, ? in
K = coefficient of thermal expansion (from Table 4-2, K for brick = 3.4×10^{-6}), 0.0000034 in/in·°F
T = temperature change in the wall (−30 to 180°F), 210°

$$1 = L \times 0.0000034 \times 210$$
$$L = 1400 \text{ in} = \frac{1400}{12} = 117 \text{ ft}$$

Number of expansion joints in wall:

$$\frac{\text{Total wall length}}{\text{Wall length per expansion joint}} = \text{number of expansion joints in wall}$$

$$\frac{320}{117} = 2.74 \qquad \text{Use 2 plus ends}$$

At least two expansion joints are needed, plus one at each end if the wall abuts other architecture.

WATER

All water coming into contact with the outside of a building should be drawn away as quickly as possible; otherwise, the liquid greatly endangers the longevity of the architecture.

Hydrostatic Head

The 24 x 40 ft concrete basement of a new residence will exist below the surrounding water table during part of the year. During this time, what will be the total pressure on the underside of the basement if no foundation drain is laid?

$$P = Ap$$

P = total pressure on the underside of the basement floor, ? lb
A = area of the basement floor, $24 \times 40 = 960 \text{ ft}^2$
p = unit pressure of hydrostatic water load, 240 lb/ft^2

$$P = 960 \times 240$$
$$= 230,000 \text{ lb} = 115 \text{ tons}$$

Foundation Drain Size

A 420-ft long shopping center wall has a basement on the inside and an excavation in moist soil on the outside. Pumping tests indicate that the quantity of water infiltrating from the excavation is 145 gal/min. Assuming the footing drain outfall is properly designed at a pitch of ⅛ in per foot, if the bottom of the footing drain is 12 in below the underside of the basement floor on the other side of the wall, what should be the minimum diameter of the footing drain? (See Fig. 4-4.)

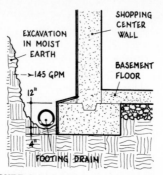

FIGURE 4-4 Footing and foundation drain.

$$7600 V d^5 = L^2 A^{1.875}$$

V = vertical dimension between the underside of the basement floor and the underside of the footing drain, 12 in

d = minimum diameter of the footing drain, ? in

L = length of the footing drain laid along the wall, 420 ft

A = infiltration rate of the water, 145 gal/min

$$7600 \times 12 \times d^5 = 420^2 (145)^{1.875}$$
$$d^5 = 21{,}800$$
$$d = 7.38 \text{ in}$$

Use 8-in diameter.

Gutter Size: Sloping Roof

In northern Florida, a house with a gable roof totalling 1700 ft^2 in area is to have semicircular gutters mounted along both lower edges. If the two gutters have six leaders, one at each corner and one in the middle of each run, how wide should the gutters be?

$$11W = RA^{0.3}$$

W = width of gutter, ? in
R = maximum rainfall intensity (from Fig. 4-5, intensity in northern Florida), 10 in/hr
A = area drained by one leader, $1700 \div 6 = 283$ ft^2

$$11W = 10 \times 283^{0.3}$$
$$= 4.94 \text{ in}$$

Use 5 in as the gutter width.

NOTE: This formula may be used for nonsemicircular gutters by adjusting the number 11 on the left side of the equation according to:

$$11 \times \frac{\text{cross-sectional area of gutter of width } D}{\text{area of semicircle of diameter } D}$$

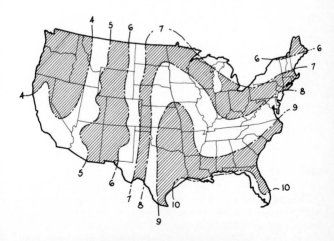

FIGURE 4-5 Rainfall intensity map.

With a standard rectangular gutter that has an ogee outer profile, use:

$$12W = RA^{0.3}$$

Leader Size: Flat or Sloping Roof

In Denver, Colorado, an office building is to be constructed with a flat roof 54 x 128 ft in size. The roof is to have four storm drains. What should the diameters of the drains be?

$$143L = RA^{0.75}$$

L = cross-sectional area of one leader, ? in^2
R = maximum rainfall intensity (from Fig. 4-5, intensity for Denver), 6 in/hr.
A = area drained by one leader, $54 \times 128 \div 4 = 1728$ ft^2

$$143L = 6 \times 1728^{0.75}$$
$$L = 11.6 \ in^2$$

Will the leader be round or rectangular? If round, find the diameter from the formula $A = 0.78D^2$. If rectangular, design the cross section according to the formula $A = BD$.

NOTE: When using this formula for sloping roofs, use the area of the horizontal projection of the roof as the roof area.

Storm Drain Size

Example 1 In the just-described Denver office building with a 54 x 128 roof, each roof drain leader becomes a horizontal (at ¼-in fall per foot) storm drain under the bottom floor. What should be the diameter of each drain? After their common junction into a single main, what should be the main's diameter?

Individual drains:

$$19D = RA^{0.38}$$

D = diameter of each individual storm drain, ? in
R = maximum rainfall intensity for Denver from Fig. 4-5, 6 in/hr
A = area drained by one leader, $54 \times 128 \div 4 = 1728$ ft^2

$$19D = 6 \times 1728^{0.38}$$
$$D = 5.37 \text{ in}$$

Use 6 in as diameter of storm drain.
   Single storm main:

$$19D = RA^{0.38}$$

D = diameter of storm main, ? in
R = maximum rainfall intensity, 6 in/hr
A = area drained, $54 \times 128 = 6910$ in^2

$$19D = 6 \times 6910^{0.38}$$
$$D = 9.09 \text{ in}$$

Use 9 in.

NOTE: When using this formula for sloping surfaces, use the horizontal projection of the surface as the drainage area.

Example 2 A swale approximately 480 ft long by an average 55 ft wide alongside a large shopping center parking lot near Baltimore, Maryland, requires drainage via a culvert. What diameter should the culvert be?

$$19D = RA^{0.38}$$

D = diameter of culvert, ? in
R = maximum rainfall intensity (from Fig. 4-5, maximum intensity for Baltimore), about 8 in/hr
A = area drained, $480 \times 55 = 26,400$ ft^2

$$19D = 8 \times 26,400^{0.38}$$
$$D = 20.2 \text{ in}$$

Use 20 in as diameter of culvert.

NOTE: This formula may be used for pavement, ground surfaces, watersheds, and for calculating the size and location of street gutters.

Drywell Size

A six-unit apartment built near St. Louis has a 2600-ft^2 roof and four leaders draining into four drywells located out from each corner. If each leader drains the same amount of roof and the perc tests on the property are 4 min per inch drop, what is the size of each drywell?

$$ARP = 1450DH$$

A = area drained by each leader, $2600 \div 4 = 650$ ft^2
R = maximum rainfall intensity (from Fig. 4-5, maximum intensity for St. Louis), about 8.5 in/hr
P = perc test result, 4 min per inch drop
D = diameter of drywell; try 3 ft
H = height of drywell, ? ft

$$650 \times 8.5 \times 4 = 1450 \times 3 \times H$$
$$H = 5.08 \text{ ft.}$$

Each drywell is a cylinder 3 ft in diameter and 5 ft high.

NOTE: The diameter dimension is to the inner face of construction. The bottom of the drywell must be at least 2 ft above the area's water table.

Cistern Size

A family of six living on a farm in eastern Nebraska plans to build a cistern for household use. If the length of time between heavy rains may be as long as 6 weeks, what should be the size of the cistern?

$$V = 4.00NL$$

V = volume of cistern, $?$ ft^3
N = number of occupants, 6
L = length of time between heavy rains in days, 6 weeks = 42 days

$$V = 4.00 \times 6 \times 42$$
$$= 1008 \text{ ft}^3$$

The cistern could be 10 ft deep, 10 ft wide, and 10 ft long.

WIND

Wind loads constitute perhaps the strongest oft-repeated climatic force that plays on architecture. Strong winds can lift a building's roof, cave in its walls, slide the structure off its foundation, and even turn the

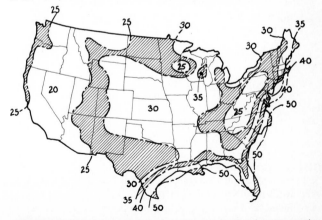

FIGURE 4-6 Wind pressure map. In some areas local codes may require different wind loads than those shown.

building over. The structural matrix of each building must be designed to withstand these forces.

Figure 4-6 is a map that shows maximum wind pressures in the 48 conterminous states.

Relation between Speed and Pressure

An 85-mi/hr wind exerts how many pounds pressure per square foot?

$$P = 0.004V^2$$

P = wind pressure, ? lb/ft^2
V = wind velocity, 85 mi/hr

$$P = 0.004 \times 85^2$$
$$= 28.9 \text{ lb/ft}^2$$

NOTE: 1 mi/hr equals 1.47 ft/sec.

Building Drift

An office tower is 440 ft tall. How much will its cornice sway in the strongest of winds?

$$D = 0.002H$$

D = probable drift, or horizontal movement, of the top of the building, ? ft
H = height of the building, 440 ft

$$D = 0.002 \times 440$$
$$= 0.88 \text{ feet, or about } 10\% \text{ in}$$

Ratio of Glass Area to Thickness

The front windows of a supermarket are 8 ft 6 in high by 10 ft 6 in wide. If the maximum local windspeed is 85 mi/hr, how thick should the windows be?

$$\text{Area of glass} = 8.5 \times 10.5$$
$$= 89.3 \text{ ft}^2$$

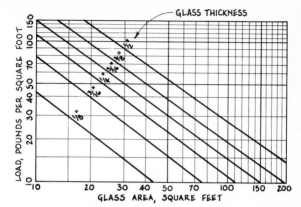

LOADS APPLY TO VERTICAL PANES SUPPORTED ON 4 SIDES AND HAVING A
LENGTH-TO-WIDTH RATIO OF 3 OR LESS.

CORRECTION FACTORS: MULTIPLY ACTUAL WIND LOAD PSF BY FACTOR BE-
LOW TO OBTAIN WIND LOAD USED IN GRAPH

TEMPERED, HEAT-STRENGTHENED GLASS	0.5
DOUBLE GLAZING (FOR THICKNESS, USE THINNER LIGHT)	0.67
FLOAT, SHEET, PLATE GLASS	1.0
LAMINATED GLASS, TOTAL THICKNESS	1.0
SANDBLASTED GLASS	2.5

FIGURE 4-7 Wind loads on graph. *(Reprinted with permission of the Architectural Aluminum Manufacturers Association, Chicago.)*

Consult Fig. 4-7. Locate intersection of horizontal "glass area" line (90 ft²) and vertical "wind velocity" line (95 mi/hr). Note the first diagonal line below this point. It is ⅜ in, but the ½-in line is very near. To be safe, use ½-in thick glass.

Horizontal Wind Pressure

A colonial-style wood frame residence near Boston is 44 ft long, 20 ft tall from foundation to eave, and the roof at its peak rises 6 ft higher.

If the strongest of winds blows directly toward the front facade, what is the total pressure against it? What architectural remedies should be taken to resist this force? (See Fig. 4-8.)

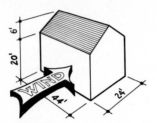

FIGURE 4-8 Wind on house.

$$W = Ap$$

W = total wind pressure against the side of the house and vertical projection of the roof, ? lb

A = total area of the facade (including all projections, such as chimneys, cupolas, bay windows on the side, etc.), $44 \times (20 + 6) = 1144 \ ft^2$

p = maximum wind pressure for the region (from Fig. 4-6, p for the Boston area), $35 \ lb/ft^2$

$$W = 1144 \times 35$$
$$= 40,000 \ lb$$

REMEDIES The wind force (W) acts upon the house in at least four ways, all of which require structural resistance.

1. The wind force tries to push the front wall down (Fig. 4-9). This creates a moment arm at A and other corners. This moment equals $W \times H/2 = 40,000 \times 26/2 = 520,000 \ ft \cdot lb$. This force is resisted by the abutting interior and end wall partitions

that are sheathed or diagonally braced. In the worst possible situation, a building shaped like the one shown in Fig. 4-10, which has no abutting interior or end partitions, the wind force must be resisted entirely at the corners A, B, C, and D.

FIGURE 4-9
Wind on house front wall.

FIGURE 4-10
Corners ABCD on house.

2. The wind force tries to distort the end walls holding the front walls in place and make them fall (Fig. 4-11). This force is resisted by plywood sheathing that is thick enough and contains enough nails hammered into the walls' framing. If the wind force is 40,000 lb, 20,000 lb is absorbed by each side. If the walls are of ½-in sheathing with 6d common nails into Douglas fir studs, each nail has a lateral resistance strength of 63 lb (from Table 3-12); each side wall needs 20,000/63 = 317 nails to provide total resistance.

FIGURE 4-11 Wind distorting end walls.

3. The wind force tries to slide the house off its foundation (Fig. 4-12). This is resisted by the weight of the house and the lateral resistance of the anchor bolts holding the bottom plate of the framing to the top of the masonry foundation wall. If the wind

force is 40,000 lb, the house weighs 80,000 lb, and the coefficient of static friction between the wood sill and concrete foundation is 0.35, the force resisted by the anchor bolts is 40,000 − 80,000 × 0.35 = 12,000 lb. If a ½-in diameter anchor bolt in concrete has an allowable shear of 550 lb (from Table 3-28), the number of anchor bolts required to keep the foundation from sliding is 12,000/550 = a minimum of 22 bolts inbedded evenly around the foundation.

FIGURE 4-12 Wind sliding house off foundation.

4. The wind force tries to turn the house over (Fig. 4-13). This is resisted by the dead weight of the house. In this case, the dwelling is considerably stronger than the wind force. However, in buildings whose ratio of height to depth is great, much tensile resistance must exist between the lower edge of the wall and the foundation. Because the wind can blow from other directions, the same resistance must exist along all other lower edges.

TENSILE RESISTANCE
FIGURE 4-13 Wind turning house over.

Corner Strength

A small barn near Boston is 20 ft tall, 24 ft wide, and has no interior partitions of any kind (Fig. 4-14). The exterior walls are vertical planking that offers no diagonal resistance. Thus, the wind force must be absorbed at corners A, B, C, and D. If the total wind force equals 40,000 lb, how should the corners of the barn be designed?

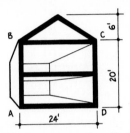

FIGURE 4-14   Barn cross section with dimensions.

STEP 1.   Half the force is on each side: 40,000/2 equals 20,000 lb on each side.

STEP 2.   One solution: install four diagonal compression braces at corners A, F, C, and H at each end (see Fig. 4-15). (Braces are also installed at corners E, B, G, and D to resist winds blowing from the other side). Design each brace to absorb 20,000/4 = 5000 lb.

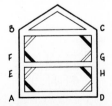

FIGURE 4-15   Barn corner braces.

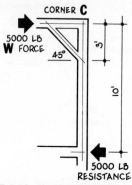

FIGURE 4-16   Barn corner brace with forces.

STEP 3.   Design the braces. Try locating them 3 ft from the corners, then design the cross-sectional area of the brace. If the size seems too heavy, try locating the brace ends farther from the corner (see Fig. 4-16).

Maximum moment at corner:

$$5000 \times 10 = 50,000 \text{ ft/lb}$$

Force resisted by brace:

$$F = \frac{M}{x} \sqrt{2}$$

F = force resisted by the 45° diagonal brace, ? lb
M = maximum moment at corner, 50,000 ft/lb
x = moment arm of the brace, 3 ft
$\sqrt{2}$ = 45° component of the diagonal brace, 1.414

$$F = \frac{50,000 \times 1.414}{3}$$
$$= 23,600 \text{ lb}$$

Cross-sectional area of brace (use Douglas fir lumber):

$$F = fA$$

F = force resisted by the diagonal brace, 23,600 lb
f = allowable compression stress of Douglas fir (from Table 3-8), 1500 lb/ft²
A = cross-sectional area of brace, ? in²

$$23,600 = 1500A$$
$$A = 15.7 \text{ in}^2$$

Use 4 × 6 (C-S area = 19.3 in²).

Seam Shear

A warehouse near Miami has concrete block walls, a flat roof of long-span steel joists, and tar and gravel roofing. If its dimensions are as shown in Fig. 4-17, in the strongest of winds what is the maximum shear along AB?

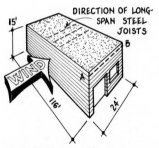

FIGURE 4-17 Seam strength.

STEP 1. Determine the total wind force against the side of the warehouse.

$$W = Ap$$

W = total wind force against the building, ? lb
A = surface area of the facade and all projections, $15 \times 116 = 1740$
   ft^2
p = design wind pressure for Miami (from Fig. 4-6, p for Miami
   area), 50 lb/ft^2

$$W = 1740 \times 50 = 87,000 \text{ lb}$$

STEP 2.   Determine the seam shear along AB.

$$v_s = \frac{W}{2L}$$

v_s = actual seam shear along meeting of roof to wall at AB, ? lb/ft^2
W = total wind force acting on the seam (two seams, one on either
   end), $87,000/2 = 43,500$ lb
L = length of seam, 24 ft

$$v_s = \frac{43,500}{2 \times 24}$$
$$= 906 \text{ lb/ft}^2$$

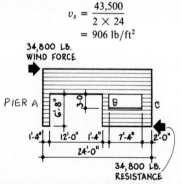

FIGURE 4-18   Warehouse front with dimensions.

Pier Rigidity

In the warehouse of the preceding problem, one end wall is designed as
shown in Fig. 4-18. The construction is grade A cinder block 8 in thick

and type N mortar. What are the pier stresses in this wall? Is the wall safe?

STRATEGY The lateral load is distributed among piers A, B, and C according to their rigidities. Each pier's rigidity depends on its width/depth ratio as expressed in the formula below.

STEP 1. Calculate the rigidity of each pier.

$$R_n = \frac{20}{3H/D + (H/D)^3} = \frac{1}{\Delta} \qquad \Delta = \text{unit deflection}$$

R = rigidity of each pier (equals the reciprocal of the deflection in the pier caused by the horizontal load)

H = height of each pier: H_a = 6 ft 8 in = 80 in, H_b = 3 ft = 36 in, H_c = 3 ft = 36 in

D = depth of each pier: D_a = 1 ft 4 in = 16 in, D_b = 1 ft 4 in = 16 in, D_c = 2 ft = 24 in

Pier A:

$$R_a = \frac{20}{(3 \times 80)/16 + (80/16)^3} = 0.143 = \frac{1}{\Delta}$$

Pier B:

$$R_b = \frac{20}{(3 \times 36)/16 + (36/16)^3} = 1.10 = \frac{1}{\Delta}$$

Pier C:

$$R_c = \frac{20}{(3 \times 36)/24 + (36/24)^3} = 2.54 = \frac{1}{\Delta}$$

STEP 2. Determine the horizontal force among the three piers according to their relative rigidities. Remember, the wind force W = 34,800 lb.

$$W_a = W\left(\frac{R_a}{R_a + R_b + R_c}\right) = 34{,}800\left(\frac{0.143}{3.78}\right) = 1320 \text{ lb}$$

$$W_b = W\left(\frac{R_b}{R_a + R_b + R_c}\right) = 34{,}800\left(\frac{1.10}{3.78}\right) = 10{,}100 \text{ lb}$$

$$W_c = W\left(\frac{R_c}{R_a + R_b + R_c}\right) = 34{,}800\left(\frac{2.54}{3.78}\right) = 23{,}400 \text{ lb}$$

STEP 3.   Check the cross-sectional area of the piers for shear. The dotted lines in Fig. 4-19 show the area of maximum shear.

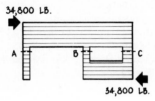

FIGURE 4-19   Warehouse front showing piers.

$$v = \frac{W}{A}$$

v = actual unit shear in the piers, ? lb/in^2
W = amount of wind force creating total shear stress, 34,800 lb
A = cross-sectional area of the piers:

> Area of pier A = $16 \times 7.5 = 120 \text{ in}^2$
> Area of pier B = $16 \times 7.5 = 120 \text{ in}^2$
> Area of pier C = $24 \times 7.5 = 180 \text{ in}^2$
> Total cross-sectional area = 420 in^2

$$v = \frac{34{,}800}{420} = 83 \text{ lb/in}^2$$

Compare this actual unit shear with the allowable shear for grade A cinder block and type N mortar from Table 3-27: allowable v for hollow units = 19 lb/in². This wall is NG in shear.

STEP 4.   Check each pier for deflection. Allowable deflection equals 1/240 of the span, which in this case is the vertical dimension of the pier. Actual deflection equals the reciprocal of the rigidity. The minimum actual value governs.

Pier A:

$$\text{Actual } \Delta = \frac{1}{0.143} = 6.99 \text{ in}$$

$$\text{Allow. } \Delta = \frac{L}{240} = \frac{80}{240} = 0.33 \text{ in} \qquad \text{NG}$$

Pier B:

$$\text{Actual } \Delta = \frac{1}{1.10} = 0.91 \text{ in}$$

$$\text{Allow. } \Delta = \frac{L}{240} = \frac{36}{240} = 0.15 \text{ in} \qquad \text{NG}$$

Pier C:

$$\text{Actual } \Delta = \frac{1}{2.54} = 0.39 \text{ in} \qquad \text{minimum value}$$

$$\text{Allow. } \Delta = \frac{L}{240} = \frac{36}{240} = 0.15 \qquad \text{NG}$$

Summary   This wall is not safe, either in shear or deflection. *Remedy:* Design the wall as shown in Fig. 4-20. Now the pier rigidities according to the above operations are $R_a = 0.42$ and R_b and $R_c = 7.48$, and the actual unit shear = 39 lb/in². If the blocks are grouted to above the openings, the allowable shear rises to 56 lb/in² and the wall becomes safe in shear. As for deflection, at the above rigidities these calculations are:

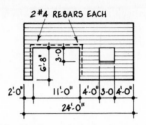

FIGURE 4-20 Warehouse front showing redesigned piers.

Pier A:

$$\text{Actual} = 2.38$$
$$\text{Allowable} = 0.33 \quad \text{NG}$$

Piers B and C:

$$\text{Actual} = 0.13 \quad \text{minimum value}$$
$$\text{Allowable} = 0.15 \quad \text{OK}$$

If one pier is OK in Δ, the wall is safe. However, it is wise to add reinforcing to all piers whose deflection is NG. Add two #4 rebars to the sides and top of the large opening.

Uplift Force

A 20 × 20 ft picnic pavilion stands on a knoll in a park near Denver, Colorado. The low-pitched roof is supported by four large posts as shown in Fig. 4-21. At each corner, the roof beams are fastened to the tops of the Douglas fir posts with four 16d toenails. If the roof weighs 4200 lb, is this condition enough to resist the maximum uplifting force in the strongest of winds, or are special connections required to hold the beam ends to the post?

STEP 1. Calculate the uplift source on the roof.

$$W = 1.25Ap$$

W = total wind uplift pressure on the pavilion roof, ? lb
A = area of horizontal plane of roof, $20 \times 20 = 400$ ft^2
p = design wind pressure for Denver (from Fig. 4-6, p for Denver area), 30 lb/ft^2

$$W = 1.25 \times 400 \times 30 = 15{,}000 \text{ lb}$$

FIGURE 4-21 Picnic pavilion.

STEP 2. Calculate the resisting strength of the roof

Resisting strength of roof = weight of roof + strength of connections

Weight of roof, 4200 lb
Nail strength at connections (from Table 3-13, page 107):
    16d nail in Douglas fir, 107 lb per nail
    Nails driven into end grain, $0.67 \times$ above
    Nails driven at angle (toe-nails), $0.83 \times$ above
    Safe withdrawal load, $0.33 \times$ above per inch penetration
    Penetration (assume 2 in), $2 \times$ above
    4 nails into each post, $4 \times$ above
    4 posts, $4 \times$ above
    Total nail resistance = $107 \times 0.67 \times 0.83 \times 0.33 \times 2 \times 4 \times 4 = 628$ lb

    Resisting strength of roof = $4200 + 628 = 4830$ lb

STEP 3. Compare the uplifting force with the resisting strength.

$$\text{Uplift source} \le \text{resistance}$$
$$15,000 \nleq 4830 \qquad \text{NG}$$

REMEDY Additional bolted plate connections that provide at least $(15,000 - 4830)/4 = 2540$ lb additional hold-down power at each post are required. Considering that the pavilion is on a knoll, which increases intensity of local winds, it would be wise to multiply this amount by 1.25.

EARTHQUAKE

Earthquakes are rare but cataclysmic in nature (see Fig. 4-22). Buildings cannot be designed to remain undamaged from such occurrences, but every effort must be made to design them to prevent their collapsing so that occupants can have time to escape.

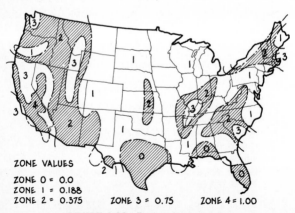

ZONE VALUES

ZONE 0 = 0.0
ZONE 1 = 0.188
ZONE 2 = 0.375   ZONE 3 = 0.75   ZONE 4 = 1.00

FIGURE 4-22 Earthquake zone map.

In earthquake analysis, seam shear forces act on structure aligned parallel to the seismic thrust, overturning forces act on structure aligned perpendicular to the seismic thrust, and torsion causes a building or part thereof to twist about its vertical axis.

Table 4-3 presents structural K factors for seismic loads.

TABLE 4-3 STRUCTURAL K FACTORS FOR SEISMIC LOADS

Type of architecture	K value
Box Systems: Shear walls or braced frames that do not exist in a complete vertical load-carrying space frame (see Fig. 4-23 below)	1.33
Dual Braced Frames: Any combination of shear walls, braced frames, and ductile frames with ductile frames absorbing a minimum of 25% of lateral force; strength of rigid elements surrounding or adjoining structural system is not included in system strength (see Fig. 4-23 below)	0.80
Ductile Moment-Resisting Space Frame: Entirely ductile frame; strength of rigid elements surrounding or adjoining structural system is not included in system strength (see below).	0.67
Elevated Tanks plus Full Contents	2.50
Structures Other Than Buildings	2.00
All Other Building Framing Systems (see Fig. 4-23).	1.00

STUD WALL W/ PLYWOOD SHEATHING   MASONRY   REIN-FORCED CONCRETE

SHEAR WALLS

FIGURE 4-23 Building framing systems.

TABLE 4-3 STRUCTURAL K FACTORS FOR SEISMIC LOADS *(Continued)*

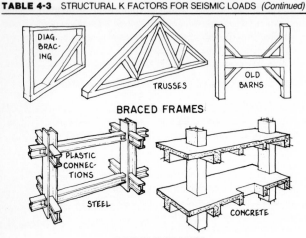

BRACED FRAMES

DUCTILE FRAMES
FIGURE 4-23 *(Continued)*

*Special conditions for steel
ductile frames*

1. Ultimate stress / Yield stress not
   greater than 1.5
2. Special precautions taken to
   prevent buckling
3. Slenderness ratios of columns
   ignores effect of adjoining shear
   walls or braced frames

*Special conditions for concrete
ductile frames*

1. Stirrups throughout lengths of
   beams
2. In beams, first stirrup is 2 in
   from joint face, next six stirrups
   are max. $d/6$ apart
3. In columns, first tie or hoop is 2
   in from joint face, next ties are
   max. 4 in apart to min. ⅙ clear
   column height from joints
4. Extra reinforcing at openings,
   joints, rebar splices, and rebar
   anchors

FIGURE 4-24   Three-story building.

Seam Shear

A three-story office building in Charleston, South Carolina, is dimensioned as shown in Fig. 4-24. Its structure is steel frame with two bays across the 52-ft span; its floors and roof are floor tile on concrete on metal decking; and its facades are 4-in brick veneer with plaster as an interior finish. Interior walls are metal studs with plaster on both sides. If an earthquake acts parallel to the side wall, what is the seismic seam shear load at the level of the second floor? This is denoted by the dotted line AB.

$$V = ZIKSCW$$

V = total seismic force acting at the level of the second floor, ? lb
Z = seismic zone factor, depending on location of the building (as shown in Fig. 4-22, Z for Charleston, S.C. area), 0.75
I = importance factor, as described below:

   Hospitals, fire and police stations, local government and communication centers, $I = 1.5$
   Buildings used for assembly of more than 300 persons in one room, $I = 1.25$
   All other buildings, $I = 1.0$
In this case, $I = 1.0$

K = structural system factor (the structural steel frame and the walls of masonry veneer act together to resist the seam shear load, thus it is a dual braced frame; from Table 4-3, K for dual braced frame), 0.80

SC = acceleration factor and subsoil factor, 0.14

W = weight of building resting on the seam; with two spans across the 52-ft dimension, count only the weight in 13 ft (52 ÷ 4) from the seam; weight equals dead load + 0.25 maximum live load; count the roof, third floor, second floor, all walls, and all other dead or live loads; obtain dead load for structural types from Table 3-1 and the live load for occupancy types from Table 3-2: Weight of near 13 ft of building: $L \times W \times (DL + 0.25LL)$

$$
\begin{aligned}
\text{Roof} &= L \times W \times (DL + 0.25LL) \\
&= 75 \times 13 \times (73 + 10) &&= 80.9 \text{ kips} \\
\text{3d floor} &= L \times W \times (DL + 0.25LL) \\
&= 75 \times 13 \times (77 + 12.5) &&= 87.3 \\
\text{2d floor} &= L \times W \times (DL + 0.25LL) \\
&= 75 \times 13 \times (77 + 12.5) &&= 87.3 \\
\text{Exterior walls} &= H \times L \times (DL + 0.25LL) \\
&= (38 - 12)(75 + 13 \times 2)(63 + 5) &&= 178.6
\end{aligned}
$$

Interior walls (err on the side of safety; assume 200 lin ft) = $H \times L \times (DL + 0.25LL)$
$$= (38 - 12)(200/4)(19 + 5) = 31.2$$

Total weight from roof to second floor   = 465.0 kips

$$
\begin{aligned}
V &= 0.75 \times 1.0 \times 0.80 \times 0.14 \times 465 \\
&= 39.1 \text{ kips}
\end{aligned}
$$

Overturning Force

An armory near Minneapolis is dimensioned as shown in Fig. 4-25. Its roof is longspan steel joists and the walls are 1-ft 8-in thick concrete with 4-in brick veneer. If an earthquake acts parallel to the side wall, what is the overturning moment per linear foot along the front edge of the roof? f_c of the reinforced concrete wall = 3000 lb/in², roof dead load = 20 lb/ft², roof live load = 30 lb/ft², wall dead load = 290 lb/ft² surface area, wall live load = 0. Is the wall safe?

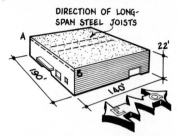

DIRECTION OF LONG-SPAN STEEL JOISTS

A

22'

190'

140'

FIGURE 4-25 One-story building.

STRATEGY Overturning force occurs when the maximum moment gathers in the roof diaphragm resting on the wall (the near half) and the part of the wall reacting against the roof (the upper half), and acts on the common seam.

STEP 1. Compute the maximum moment at the seam of roof to wall AB.

$$M = \frac{WL}{48}$$

M = maximum moment at the seam of roof to wall due to seismic force, ? ft · lb

W = weight of front half of roof and upper half of wall, $W = D + 0.25L$:

   Weight of roof, $190 \times 70(20 + 7.5) = 366$ kips
   <u>Weight of wall, $190 \times 11(290 + 5) = 617$ kips</u>

   Total weight acting upon the seam = 983 kips

L = length of seam of roof to wall, 190 ft

$$M = \frac{983 \times 190}{48} = 3890 \text{ kip} \cdot \text{ft}$$

STEP 2. Compute the minimum thickness of the reinforced concrete wall holding up the joists and compare it to the wall's actual thickness.

$$M = fS$$

M = maximum moment at the seam of wall to roof in inch pounds, 3890 kip·ft $\times$ 12,000 = 46,200,000 in·lb

f = allowable concrete unit stress, 3000 lb/in^2

S = section modulus of wall cross section (cross section is a long narrow rectangle), $S = bd^2/6$:

   b = length of wall, 190 $\times$ 12 = 2280 in

   d = thickness of wall, ? in

$$S = \frac{2280d^2}{6} = 380d^2$$

$$46,700,000 = 3000 \times 380d^2$$
$$d = 6.36 \text{ in minimum}$$

Actual thickness of concrete:

$$24 \text{ in} - 4\text{-in veneer} = 20 \text{ in} > 6.40 \text{ in} \qquad \text{OK}$$

Torsion

In the three-story office building located in Charleston, South Carolina, from "Seam Shear" above, what are its torsion stresses at ground floor level due to seismic loads? A structural plan of the building is shown in Fig. 4-26. All 12 columns are W 8 $\times$ 31 A36 steel.

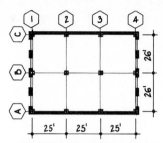

FIGURE 4-26 Building plan.

STRATEGY Torsion stress stems from the twisting of the building about its vertical axis due to rotational seismic action. The stress acts as a horizontal shear through the cross-sectional area of the columns or load-bearing walls on each floor.

$$v_t = \frac{W(0.05 + e)\sqrt{b^2 + d^2}}{bKA_s}$$

v_t = unit shear stress due to torsion (this must be less than the allowable shear; if not, the cross-sectional area of the 12 columns must be made larger), ? kips/in^2

W = weight of building above the shear plane, in this case the weight of the building above the ground floor level; count the roof, third and second floors, exterior walls from ground floor up, and non-load-bearing interior walls from second floor up; weight = dead load + 0.25 live load:

$$
\begin{aligned}
\text{Roof} &= L \times W \times (\text{DL} + 0.25\text{LL}) \\
&= 75 \times 52 \times (73 + 10) &= 324 \text{ kips} \\
\text{3d floor} &= L \times W \times (\text{DL} + 0.25\text{LL}) \\
&= 75 \times 52 \times (77 \times 12.5) &= 349 \\
\text{2d floor} &= L \times W \times (\text{DL} + 0.25\text{LL}) \\
&= 75 \times 52 \times (77 \times 12.5) &= 349
\end{aligned}
$$

$$\text{Exterior walls} = L \times W \times (\text{DL} + 0.25\text{LL})$$
$$= 2(75 + 52) \times 38 \times (63 + 5) = 656$$

Interior load-bearing walls (assume all
interior walls are non-load-bearing; 200 lin ft
interior walls per floor) $= L \times H \times (\text{DL} + 0.25\text{LL})$
$$= 200 \times (38 - 12) \times (19 + 5) = 125$$

Total weight of part of building above shear plane = 1800 kips

e = eccentricity due to nonrectangular shape; shape of building is rectangular, thus $e = 0$ (see Note below)

b = shorter dimension of rectangle, 52 ft (see Note below)

d = longer dimension of rectangle, 75 ft (see Note below)

K = d/b ratio as listed below (interpolate for intermediate values):

d/b	K	d/b	K
1.0	0.208	4.0	0.282
1.5	0.231	5.0	0.291
2.0	0.246	7.0	0.302
2.5	0.258	10.0	0.312
3.0	0.267	∞	0.333

$d/b = 75/52 = 1.44$ $K = 0.23$

A_s = cross-sectional area of the columns or load-bearing walls on each floor: C-S area of W 8 × 31 = 9.13 in², C-S area of all 12 columns = 9.13 × 12 = 110 in²

$$v_t = \frac{1800(0.05 + 0)\sqrt{52^2 + 75^2}}{52 \times 0.23 \times 110} = 6.24 \text{ kips/in}^2$$

Allowable v_t = allowable unit shear for A36 steel = 14.5 kips/in²; 14.5 kips/in² > 6.24 kips/in²   OK

NOTE: For nonrectangular plans, the value of e is found according to the following procedure.

1. Draw the smallest rectangle around the plan that wholly contains the plan shape.
2. Find the centroid of moments about the X axis of both the actual plan and the superimposed rectangle.
3. Compute the distance between the two centroids along the X axis.
4. Find the centroid of moments about the Y axis of both the actual plan and the superimposed rectangle.
5. Compute the distance between the two centroids along the Y axis.
6. Find the hypotenuse of the right triangle formed by the lengths found in steps 3 and 5.
7. Find the hypotenuse of the superimposed rectangle.
8. Find *e*. It equals the length found in step 6 divided by the length found in step 7.

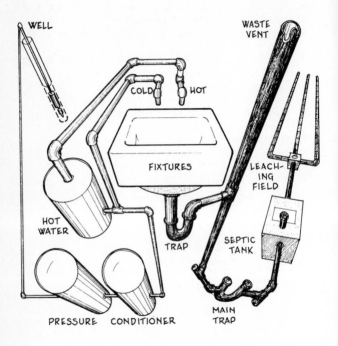

Plumbing system.

1 2 3 4 **5**
6 7 8 9
0 1 10

PLUMBING

GENERAL

The most important determinant in the sizing of plumbing components is the number of fixture units in the architecture. By looking at the plans and adding up the fixture unit numbers according to Table 5-1, a base number can be obtained that may be used to size many parts of a plumbing system.

TABLE 5-1 FIXTURE UNIT SPECIFICATIONS

Fixture	Fixture units, private	Fixture units, public	Min. flow pressure, lb/in²
Urinal:			
Flush tank	—	5	15
Flush valve	—	10	15
Toilet:			
Flush tank	3	5	8
Flush valve	6	10	15
Bidet	1	—	8
Lavatory:			
Standard	1	2	8
Self-closing faucet	1	2	12
Bathtub:			
No shower	2	4	8
With shower	2	4	12
Stall shower	2	2	12
Bathroom unit:			
Flush tank	6		
Flush valve	8		
Kitchen sink	2	4	8
Dishwasher	2	4	8
Bar sink	1	3	8
Clothes washer	2	4	8

TABLE 5-1   FIXTURE UNIT SPECIFICATIONS (Continued)

Fixture	Fixture units, private	Fixture units, public	Min. flow pressure. lb/in²
Laundry sink	3	4	8
Service sink	2	3	8
Hose bibb	2	4	30
Drinking fountain	2	2	15
Water cooler	2	2	8
Fire hose	0	0	30

A *fixture unit* is a unit flow rate approximately equal to 1 ft³/min maximum flow into or out of a plumbing fixture.

When designing plumbing systems, consult applicable local codes that may govern for a specific design.

Fixture Units

How many fixture units does the partial plan in Fig. 5-1 have?

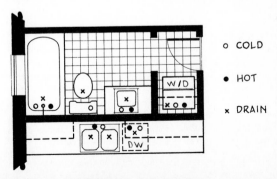

o COLD

• HOT

x DRAIN

FIGURE 5-1   Bathroom plan.

The plumbing fixtures in plan are listed below.

Fixture	*Fixture units*
Bathroom unit, flush tank	6
Washer with piggyback dryer	2
Kitchen sink	2
Dishwasher	2
Total	12

SUPPLY AND SANITARY WASTE

For the most part, drinking water enters the architecture from one part of the property and leaves through another. In meter-sewage systems the supply and drain stems may be laid in the same trench. However, in well-septic systems, the two stems should be at least 100 ft apart.

Calculating the size of plumbing components is not an exact science. As a rule, estimates include a slight safety factor.

Pipe Sizing

Example 1   A nine-unit apartment building is served by a submersible well pump and pressure tank system whose minimum pressure is 35 lb/in^2. The number of fixture units for each apartment is 14 (flush tank toilets), the length of piping from the pressure tank to the most remote fixture is 68 ft, the head (vertical elevation difference) from the pressure tank to a third-floor bathroom shower head is 23 ft, and the pressure at the fixture requiring the maximum flow pressure is 12 lb/in^2. From this information, calculate the minimum diameter of the cold water service mains to each apartment.

STEP 1.   Calculate the pipe pressure friction loss according to the following formula.

$$F = \frac{57P - 29H - 67M}{L}$$

F = pressure loss caused by pipe friction, ? lb/in² loss per 100 lin ft
P = water pressure at beginning of run (minimum), 35 lb/in²
H = head, 23 ft
M = maximum fixture pressure, 12 lb/in²
L = length of pipe (the formula includes safety factors for fittings and meter pressure drop), 68 ft

$$F = \frac{57 \times 35 - 29 \times 23 - 67 \times 12}{68} = \frac{524}{68} = 7.71 \text{ lb/in}^2$$

NOTE: This formula should not be used for runs containing an abnormally large number of bends and fittings.

STEP 2. Knowing the number of fixture units and pressure loss, find the minimum pipe diameter from Fig. 5-2.

From Fig. 5-2, minimum pipe diameter is 1-in copper tubing. Note that water velocity through the pipe is less than 8 ft/sec. If it exceeds 8 ft/sec, the pipe must be resized according to the following formula.

$$D = F\sqrt{0.125V}$$

D = pipe diameter required, in inches
F = pipe diameter selected in graph, in inches
V = velocity selected in graph, in feet per second

Example 2 A four-story office building requires a copper service pipe from the town water main out front, which has a street pressure of 50 lb/in². From the plans, the building's total number of fixture units is 176, the total length of piping from the town main to the most remote fixture is 430 ft, the head from the water main to a cooling tower spigot on the roof is 45½ ft, and this spigot is the fixture requiring the maximum flow pressure (30 lb/in²). From this information, calculate the diameter of the incoming service main.

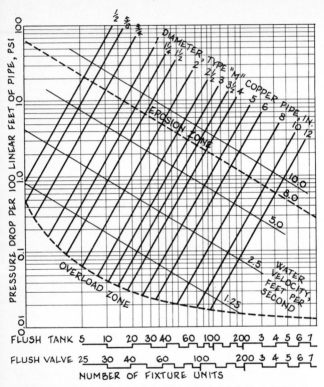

EROSION ZONE: AT VELOCITIES ABOVE UPPER DOTTED
LINE, PIPE EROSION MAY OCCUR; CHOOSE LARGER DIA-
METER PIPE.
OVERLOAD ZONE: AT VELOCITIES BELOW LOWER DOTTED
LINE, PLUMBING SYSTEM LOAD IS TOO GREAT OR INCOM-
ING PRESSURE IS TOO LOW; REDESIGN SYSTEM.

FIGURE 5-2 Pipe diameter graph.

STEP 1. Find the pipe pressure friction loss:

$$F = \frac{57P - 29H - 67M}{L}$$

F = pressure loss caused by pipe friction, ? lb/in^2
P = water pressure at beginning of run, 50 lb/in^2
H = head, 45.5 ft
M = maximum fixture pressure, 30 lb/in^2
L = length of pipe, 430 ft

$$P = \frac{57 \times 50 - 29 \times 45.5 - 67 \times 30}{430} = \frac{-480}{430} = -1.12 \text{ lb/in}^2$$

When P is a negative number, something is wrong with the system design. Usually the incoming water pressure is not high enough to overcome the fixture pressure load created by the head, pipe length, and flow pressure requirements. This is the case even when pipe pressures are positive but near zero, as when they fall below the lower dotted line in the graph of Fig. 5-2. In such situations, the water supply system must be redesigned. This is usually done by installing a pneumatic tank or tankless system pump somewhere inside the building.

Component Sizing

The formulas presented in this section are approximate and exceed minimum requirements for typical residential and commercial architecture.

Supply water delivery rate   A motel to be built on country property will obtain its water supply from a drilled well containing a submersible pump. If the architecture contains 132 fixture units, what is the minimum water flow rate that should be struck in the well?

$$\log (1.67D) = 0.77 \log F$$

D = delivery rate of supply water, ? gal/min
F_u = number of fixture units served, 132

$$\log (1.67D) = 0.77 \log 132$$
$$= 1.63$$
$$D = 25.7 \text{ gal/min}$$

NOTE: The gallon capacity of a pressure tank accompanying a well pump should be about 10 times the gallons-per-minute capacity of the pump.

This formula may be used to estimate the water usage of any architecture, whether its water be drawn from the ground, surface drainage, or municipal supply.

Hot water heater capacity A hot water heater must serve a four-unit apartment that contains 51 fixture units. How big should the heater be?

$$\log (0.33H) = 0.83 \log F$$

H = hot water heater capacity, $?$ gal
F = number of fixture units served, 51

$$\log (0.33H) = 0.83 \log 51$$
$$= 1.417$$
$$H = 79.2 \text{ gal}$$

From product information, select the smallest hot water heater whose capacity exceeds this amount.

Sanitary drain diameter An office park contains six medium-sized buildings. The sanitary drains (excluding storm drainage) pour into one sanitary main that empties into the city sewage system. If the office park contains 624 fixture units, what diameter should the sanitary main be?

STEP 1. Decide what pitch the pipe should have. From analyzing the plans, you have decided that a ¼-in fall per foot is best.

STEP 2. Knowing the pitch, use the appropriate formula below.

Pitch   *Formula*

⅛ in fall per foot $\log (1.35D) = 0.33 \log F$
¼ in fall per foot $\log (1.42D) = 0.33 \log F$
½ in fall per foot $\log (1.48D) = 0.33 \log F$

Use $\log (1.42D) = 0.33 \log F$.

D = diameter of waste drain, ? in
F = number of fixture units emptying into system (for waste plumbing
design, do not count hose bibbs and other fixtures that do not con-
nect into waste drains), 624 units

$$\log (1.42D) = 0.33 \log 624$$
$$= 0.922$$
$$D = 5.89 \text{ in}$$

Use 6 in.

NOTE: The minimum sanitary main diameter for any architecture
is 4 in.

In installations having less than 6-in diameters, a pitch of ⅛ in is not
recommended.

Septic tank size   A residence built on 1.31 acres of land will
empty its sewage into a septic tank. If the architecture contains 28 fixture units, not counting hose bibbs, what size should the septic tank be?

STEP 1. Solve for the length of the septic tank by using the formula
below.

$$4 \log L = \log (190F)$$

L = length of septic tank, ? ft
F = number of fixture units emptying into system, 28

$$4 \log L = \log (190 \times 28)$$
$$= 3.73$$
$$L = 8.54 \text{ ft}$$

STEP 2. Knowing the width of a septic tank is about 0.5 its length and the height about 0.6 its length, design the septic tank.

$$\text{If } W \approx 0.5L, \ W \approx 0.5 \times 8.54 \approx 4.3 \text{ ft}$$
$$\text{If } H \approx 0.6L, \ H \approx 0.6 \times 8.54 \approx 5.1 \text{ ft}$$

Rounding off, make $L = 9$ ft, $W = 4.5$ ft, $H = 5$ ft.

NOTE: LWH measures the interior volume of the septic tank, not its exterior size.

Leaching tile length The sewage of a duplex residence containing 41 fixture units (excluding the hose bibbs) drains into a cesspool which is no longer useful. A leaching field must now be laid from the new septic tank. If the perc test is 3 min per inch drop and the leaching trenches are 2 ft wide, what is the total length of the leaching stems?

STEP 1. Find the volume of sewage from the formula below.

$$\log V = 0.78 \log (235F)$$

V = volume of sewage per day, ? gal
F = number of fixture units emptying into system, 41

$$\log V = 0.78 \log (235 \times 41)$$
$$= 3.11$$
$$V = 1280 \text{ gal/day}$$

STEP 2. Knowing the perc test result, find the unit tile length from Table 5-2. If the perc test result is 3 min per inch drop and the leaching trench is 2 ft wide, the unit tile length is 0.18 ft/gal of sewage.

STEP 3. Multiply the volume of sewage times the unit tile length to obtain the total leaching stem length in linear feet.

$$1280 \times 0.18 = 231 \text{ lin ft of leaching stem}$$

TABLE 5-2 LEACHING TILE LENGTH PER GALLON OF SEWAGE

	Perc test: minutes per inch drop	Unit tile length in feet for trench widths of		
		1 ft	2 ft	3 ft
Porous soils	1	0.25	0.13	0.09
	2	0.30	0.15	0.10
↑	3	0.35	0.18	0.12
	5	0.42	0.21	0.14
	10	0.59	0.30	0.20
	15	0.74	0.37	0.25
	20	0.91	0.46	0.31
Impermeable soils	25	1.05	0.53	0.35
	30	1.25	0.63	0.42

Fire Sprinkler System: Area per Head

A sprinkler system is to be installed in a new office building that will have suspended ceilings. If sprinkler heads are to be mounted above and below the ceiling, what is the maximum allowable area per head? (Refer to Table 5-3.)

From Table 5-3, the maximum allowable area per sprinkler head for the office under the conditions described is 200 ft².

TABLE 5-3 SPRINKLER HEAD SPECIFICATIONS

Hazard class	Light	Ordinary	Extra
Area per sprinkler head, space above ceiling not sprinklered, ft²	168	130	90
Area per sprinkler head, space above ceiling sprinklered, ft²	200	150	105
Area per head, sprinklers above ceiling, system hydraulically designed, ft²	225	175	120
Maximum spacing between sprinkler heads, ft	15	15	12

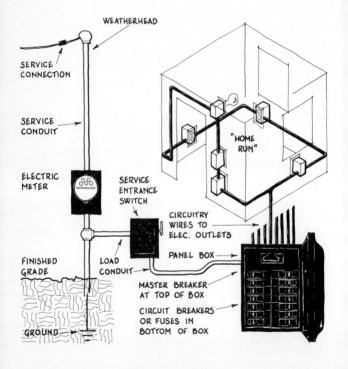

WEATHERHEAD

SERVICE
CONNECTION

SERVICE
CONDUIT

ELECTRIC
METER

SERVICE
ENTRANCE
SWITCH

"HOME
RUN"

CIRCUITRY
WIRES TO
ELEC. OUTLETS

PANEL BOX

FINISHED
GRADE

LOAD
CONDUIT

MASTER BREAKER
AT TOP OF BOX

CIRCUIT BREAKERS
OR FUSES IN
BOTTOM OF BOX

GROUND

Small electrical system.

ELECTRICITY

GENERAL

The electrical system of a building has three general parts: *service* (where the electricity originates and is regulated), *distribution* (the conduits and components that disperse the electric energy to its destination), and the *circuits* (where the electricity is consumed). After determining the location of the building's service components, the architect should size the various parts of the electrical system from its points of use backwards: that is, first size the circuits, then the distribution elements, and finally the service components.

When designing electrical systems or any part thereof, an architect should consult the National Electrical Code or applicable local codes that may govern for specific design.

LAYOUT

The size of a building's electrical system may be classed as small, medium, or large. Below are a description and diagram of each.

Small   The total system is regulated by one panel box (also known as a breaker box, circuit breaker box, or fuse box), with possibly an adjacently located panel box for electric heating. The service is almost always 120/240 V and 60, 100, or 200 A. There is rarely any three-phase wiring or large motors on individual circuits; and the lighting fixtures and electrical outlets in one space are usually on the same circuit. Distribution elements between panel box and points of use are virtually nonexistent. This size system is typical of residences and very small commercial buildings.

A small electrical system is illustrated at the opening of this chapter.

Medium   The total system is regulated by a main switchboard and several panel boxes, the latter being located in various parts of the building. The electrical service may be 200 A or more and as high as 480 V. There may be three-phase wiring, small transformers, and a large motor or two having its own circuit. Lighting fixtures and electrical outlets are on separate circuits and usually are protected by separate panel boxes. Distribution wiring, though thicker, compares in length to circuit wiring. (See Fig. 6-1.)

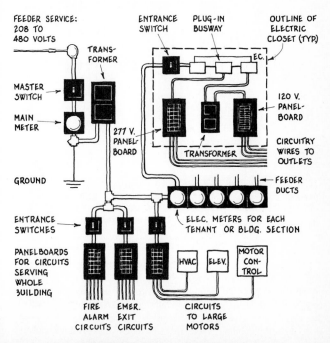

FIGURE 6-1 Medium electrical system.

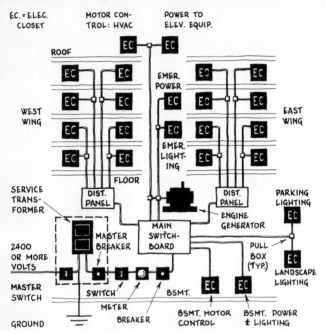

FIGURE 6-2   Large electrical system.

Large   This complicated system serves acres of floor area. The primary electrical service is high voltage (usually 2400, 4160, 7200, or 13200), often enters the building underground, and passes through a large stepdown transformer before entering any regulating equipment. Busways, entrance switches, and panel boxes exist in electrical closets throughout the building. There are large motors that operate elevators, central HVAC equipment, and industrial or business machinery. Circuit wiring is hidden above suspended ceilings, laid under access floor panels, or installed in wall-mounted raceways. Distribution systems are characterized by long feeder ducts, tall riser shafts, distribution panels, pull boxes, and stepdown transformers. (See Fig. 6-2.)

Large electrical systems are found only in skyscrapers, convention halls, corporate office buildings, and other large buildings or complexes.

SERVICE

Service components include entry facilities (overhead or underground), meters, master switches, panel boards, and grounding.

Relation between Power, Potential, and Current

How many watts are in an electrical service entrance of 240 V at 100 A?

$$W = AV$$

W = amount of electric power in watts, ? W
A = amount of electric current in amperes, 100 A
V = amount of electric potential in volts, 240 V

$$\begin{aligned} W &= 100 \times 240 \\ &= 24,000 \text{ W maximum power available} \end{aligned}$$

NOTE : 1 kW = 1000 W

Initial Estimate

An elementary school floor plan measures 28,450 ft^2 in gross area. What would be a proper size service for the building, in amperes and volts?

STEP 1. Find the estimated total wattage required by the building.

$$W = aL$$

W = total estimated electrical load of building, ? W
a = gross floor area of building, 28,450 ft^2
L = use load factor, as listed below:

Use load	Building type	Estimated W/ft^2
Very low	Parking garages, barns, buildings whose use requires little electricity	4
Low	Residences, apartments, hotels, motels	6
Low-medium	Grammar schools, athletic facilities, supermarkets, retail outlets, museums, churches, movie theatres	8
Medium	Senior high schools, universities, offices, taverns, restaurants, opera houses, bus terminals	10
Medium-high	Night clubs, luncheonettes, hospitals, spaces requiring steady round-the-clock use of electricity	12
High	Factories, industrial laboratories, buildings housing constantly used large machinery	14

An elementary school is low-medium = 8 W/ft^2.

$$W = 28,450 \times 8 = 228,000 \text{ W, or } 228 \text{ kW}$$

STEP 2. Select proper voltage from the list below.

Total use load, kW	*Voltage*
To 12: Small residences, outbuildings	120
12–96: Usual residential, small commercial	240
96–900: Most commercial buildings	208
400–2000: Large commercial-industrial	480
Specialized commercial or industrial	600
	2400
2000+: Skyscrapers, corporate offices, other large structures	4160
	7200
	13200
	34500

At 228 kW commercial, voltage = 208.

STEP 3. Find the base ampacity from the formula below.

$$W = 1.73 \; AV$$

W = total electric load of building, 228,000 W
A = ampacity of electrical load, ? A
V = voltage of electrical service, 208 V

$$228,000 = 1.73A \times 208$$
$$A = 634 \text{ A}$$

STEP 4. Select the final estimated ampacity from the numbers below. Pick the smallest number that is greater than the base ampacity.

60 100 200 400 800 1200 1600 2000 2500

The smallest number above 634 is 800. The estimated service load for the school is 208 V at 800 A.

Overhead Service Entry

What are the specifications for an overhead service entry (Fig. 6-3)?

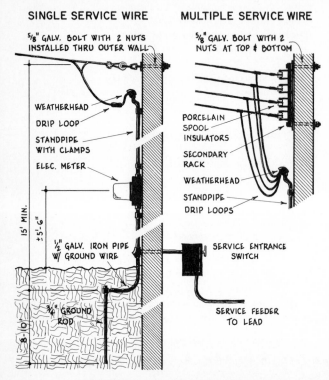

FIGURE 6-3   Overhead service entries.

Underground Service Entry

What are the specifications for an underground service entry (Fig. 6-4)?

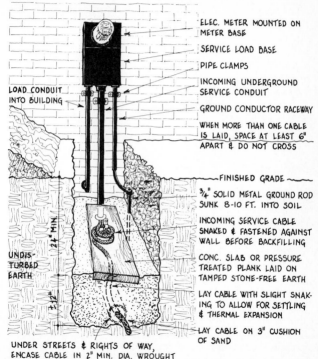

ELEC. METER MOUNTED ON METER BASE

SERVICE LOAD BASE

PIPE CLAMPS

INCOMING UNDERGROUND SERVICE CONDUIT

LOAD CONDUIT INTO BUILDING

GROUND CONDUCTOR RACEWAY

WHEN MORE THAN ONE CABLE IS LAID, SPACE AT LEAST 6" APART & DO NOT CROSS

FINISHED GRADE

3/4" SOLID METAL GROUND ROD SUNK 8-10 FT. INTO SOIL

INCOMING SERVICE CABLE SNAKED & FASTENED AGAINST WALL BEFORE BACKFILLING

24" MIN.

UNDIS-TURBED EARTH

CONC. SLAB OR PRESSURE TREATED PLANK LAID ON TAMPED STONE-FREE EARTH

LAY CABLE WITH SLIGHT SNAK-ING TO ALLOW FOR SETTLING & THERMAL EXPANSION

12"

LAY CABLE ON 3" CUSHION OF SAND

UNDER STREETS & RIGHTS OF WAY, ENCASE CABLE IN 2" MIN. DIA. WROUGHT IRON CONDUIT

FIGURE 6-4 Underground service entry.

Measurement of Electric Power

At the beginning of last month, the dials on a household electric meter read 4 2 7 1 3. At the end of the month they read 4 4 0 8 7. If the local utility charges 6.25 cents per kilowatthour, what is the month's electricity bill for the owners of the house?

$$E = (H - L)R$$

E = total electric energy used between readings, **?** dollars
H = higher of the two readings, 44,087
L = lower of the two readings, 42,713
R = electric rate charged by the utility in dollars/kWh, 6.25¢/kWh
    = \$0.0625/kWh

$$E = (44,087 - 42,713)0.0625 = \$85.88$$

CIRCUITS

Circuit design involves analyzing the plans, determining the number of electrical outlets, then joining the outlets in groups of convenient and economical voltage and amperage. If available, plans showing cabinetry and furniture layouts are helpful.

The four types of circuits are described below. They must be separate from each other, except that in residences lighting and convenience outlets in one space may be on the same circuit.

Machinery

Large motors and mechanical devices each require one circuit that supplies the electric energy to that piece of equipment only. Examples are electric ranges, refrigerators, clothes dryers, air-conditioning units, photocopiers, large computers, elevator motors, HVAC machinery, and large industrial mechanisms. On the architectural plans, such equipment should first be positioned accurately, then its circuitry can be determined by the same data that appear on its nameplate.

Example   The nameplate on an elevator motor shows the following data: 5 HP, 208 V ac, 32 A. What is the size of its circuit in amperes and volts?

$$C = 1.25A \text{ at } V$$

C = circuit ampacity at listed voltage, ? A
A = amperage of machinery as listed on its nameplate, 32 A
V = voltage of machinery as listed on its nameplate, 208 V

$$C = 1.25 \times 32$$
$$= 40 \text{ A at } 208 \text{ V}$$

Appliances

These are circuits for areas of heightened electrical use, such as kitchens, workshops, school laboratories, theatre projection rooms, and the like. In apartments and residences, at least two 20-A appliance circuits should exist for the kitchen, pantry, dining room, and family areas. There should be no more than three outlets per circuit, one outlet should be in the dining room, and one more outlet should exist in either the family or living room. In commercial architecture, appliance circuits are sized by analyzing specific requirements.

Lighting

Lighting outlets are switch-operated outlets. They are fitted either with a lighting fixture or a convenience outlet into which is plugged a lamp with cord. The lighting outlets in one space should be on at least two circuits. All lighting switches should be mounted at chest height within 2 ft of the striker side of the doorway. If a room has more than one doorway, three-way switches should be installed by each.

Unfinished areas should have one lighting outlet for every 200 ft² of floor area. Each outlet is rated at 240 W, or 2.0 A at 120 V.

Finished residential areas should have a minimum lighting load of 3 W/ft², or at least one lighting outlet in each space.

The lighting load for commercial spaces of average height and the coefficient of utilization (see section on Fundamentals under "Coefficient of Utilization" in Chap. 7) may be estimated according to the formula below. For other conditions, lighting circuitry is calculated after the fixtures have been designed.

Example What is the lighting circuitry load of a 12 x 18 ft living room? How many outlets are required?

$$AV = Wa$$

A = lighting circuitry load in amperes, ? A
V = voltage of the lighting system, assume 120 V
W = required illumination in the space, 3 W/ft^2
a = gross floor area of the space, 12 x 18 = 216 ft^2

$$A \times 120 = 3 \times 216$$
$$A = 5.4 \text{ A at } 120 \text{ V}$$
$$\text{Number of outlets} = \frac{\text{circuitry amperes}}{\text{amperes per outlet}} = \frac{5.4}{2} = 2.7$$

Three outlets are required.

Convenience

Convenience outlets are located throughout the architecture for general use by the occupants. They are 120 V. In residences, convenience outlets are measured at 2.5 A each; in commercial spaces they are 4 A each. Convenience outlet requirements in all architecture are as follows.

Finished spaces: 1 outlet per 12 lin ft of wall space for rooms up to 16 ft wide plus one outlet per 100 ft^2 of interior space more than 8 ft away from walls

Rest rooms and baths: 1 ground fault interruption (GFI) outlet in each

Corridors: 1 20-A appliance outlet per 50 ft of hall length

Unfinished areas: 1 outlet per 20 lin ft of wall space plus 2 per
1000 ft^2

Outside: 1 GFI outlet per building

Example 1 How many convenience outlets should a 21 x 40 ft
engineering drafting room have? What is the room's convenience circuitry load?

$$N = 0.083P + \frac{A}{100}$$

N = number of convenience outlets in the space, ? units

P = perimeter of room, 2(21 + 40) = 122 ft

A = area of part of room more than 8 ft away from walls (21 − 16)(40 −
16) = 5 × 24 = 120 ft^2

$$N = 0.083 \times 122 + \frac{120}{100}$$
$$= 11.3 \rightarrow 12 \text{ outlets}$$

Circuitry load = 12 outlets @ 4 A = 48 A at 120 V.

Example 2 How many convenience outlets should a 12 x 18 ft living room have? What is the room's convenience circuitry load?

For finished spaces:

$$N = 0.083P$$

For unfinished areas:

$$N = 0.05P$$

N = number of convenience outlets in the space, ? units

P = perimeter of space in feet; perimeter of the living room = 2(12
+ 18) = 60 ft

$$N = 0.083 \times 60$$
$$= 5 \text{ convenience outlets}$$

Circuitry load = 5 outlets @ 2.5 A = 12.5 A at 120 V.

Summary When the circuitry loads in each space have been calculated, the loads should be organized into circuits. Each piece of
machinery has its own. The appliance circuits can be added up in the
20-, 25-, or 30-A units in which they were designed. As for the lighting
and convenience outlets, each should be arranged into 15-, 20-, and
(sometimes in commercial architecture) 25- and 30-A groups. Each
group becomes a circuit. After arranging this wiring for each space,
make a list of all the circuits, along with the amperes, volts, and areas
served by each. This accounting is known as a *panel list*. It will be used
to size the panel boxes.

WIRE SIZES

Electrical wire is sized so that the maximum voltage drop through its
length is less than 5 percent. If the voltage drop exceeds this amount, a
smaller numbered wire (which has a larger diameter and impedes the
flow of electrons less) should be selected.

Wire sizes for the most common circuits if they are less than 100 ft
long are as follows.

Circuit	*Wire size*
15 A at 120 V	no. 14
20 A at 120 V	no. 12
30 A at 120 V	no. 10

Wiring for other ampacities and lengths may be sized as below.

Example A farmer in Florida has a shop for repairing farm
machinery 800 ft behind his residence, which has 120/240 electrical
service. The farmer wants to replace the shop's gasoline generator with
overhead electrical wiring. He needs about 12 kW to run the facility's

portable arc welder and hand-powered tools and to light the place at night. What size copper wire should he run from his house to the shop?

PROCEDURE Try different wire sizes in the formula below until you find one that has a voltage drop of less than 5 percent.

Try no. 4 wire at 120 V.

$$P_v = \frac{200RAL}{CV} \qquad \text{or} \qquad \frac{200RWL}{CV^2}$$

P_v = percentage drop in voltage of circuit (maximum allowable is 5 percent), ? %

R = resistance of conductor material in ohms per foot per cmil·ft, 12 for copper, 18 for aluminum or aluminum-clad wire; use copper wire, 12

A = ampacity of circuit if given; not given (use second formula)

W = wattage of circuit if given, 12 kW = 12,000 W

L = length of circuit from house to shop, 800 ft

C = cross-sectional area of wire in circular mils (as listed in Table 6-1 for no. 4 wire), 41,740 cmil

V = voltage of circuit (use 120 or 240), try 120 V

$$P_v = \frac{200 \times 12 \times 12,000 \times 800}{41,740 \times 120^2}$$
$$= 38.3\% \qquad \text{Much too high}$$

Try no. 1 wire at 240 V.

$$P_v = \frac{200RWL}{CV^2}$$

All values are the same except C and V.

C = cross-sectional area of no. 1 wire, 83,690 cmil

V = voltage of circuit, 240 V

$$P_v = \frac{200 \times 12 \times 12{,}000 \times 800}{83{,}690 \times 240^2}$$
$$= 4.78\% \quad \text{OK}$$

TABLE 6-1 ELECTRICAL WIRE PROPERTIES

Wire size (AWG or MCM)	C-S area, cmil*	Max. number of wires in conduit			
		Conduit diameter, in			
		½	¾	1	1¼
16	2580	9	15	25	44
14	4109	7	12	19	35
12	6530	5	9	15	26
10	10,380	2	4	7	12
8	16,510	½″ 1	2	4	7
6	26,240	1	3	5	7
4	41,740	¾″ 1	2	4	6 1½″
2	66,360	1	2	4	5
1	83,690	1	3	4	6
0 (1/0)	105,600	1″ 1	2	3	5 2″
00 (1/0)	133,100	1	3	5	7
000 (1/0)	167,800	1¼″ 1	2	4	6 2½″
0000 (1/0)	211,600	1	3	5	7
250 MCM	250,000	1½″ 1	2	4	6 3″
300 MCM	300,000	1	2	3	5
400 MCM	400,000	2″ 1	2	4	5 3½″
500 MCM	500,000	1	3	4	6 4″
		2½″	3″	3½″	4″

*A 1-in-diameter wire contains 1,000,000 cmil.

DISTRIBUTION

After all the circuits have been designed, they must be connected to the panel boxes. In small electrical systems this is simple, but in medium or large networks, thick wiring often travels in long horizontal raceways, up vertical riser shafts, into electrical closets filled with equipment, and through stepdown transformers before branching into circuits.

Electrical closets typically have about 2 x 8 ft interior floor dimensions if they have double-door access, and 4 x 5 ft floor dimensions for single-door access.

All distribution wiring, switches, panel boxes, receptacles, and other components should be jacketed in steel; then if any malfunctions occur, the enclosing metal will contain any arcs, dissipate any heat, and snuff any fire.

When sizing panel boxes, about 20 percent vacant breaker space should be left for possible future expansion.

Distribution Wiring Size

What wire size should be used for a copper feeder cable carrying 480 V at 200 A whose run from main switchboard to the most remote electrical closet is 435 ft?

$$P_v = \frac{200RAL}{CV} \quad \text{or} \quad \frac{200RWL}{CV^2}$$

Try no. 00 (1/0) wire at 480 V and 200 A

P_v = percentage drop in voltage of circuit (maximum allowable is 5 percent), ?%

R = resistance of conductor material, 12 for copper, 18 for aluminum or copper-clad aluminum wire

A = ampacity of circuit if given, 200 A
W = wattage of circuit if given; not given, so use first formula
C = cross-sectional area of wire in circular mils as listed in Table 6-1
      for no. 00 wire, 133,100 cmil
V = voltage of circuit, 480 V

$$P_v = \frac{200 \times 12 \times 200 \times 435}{133,100 \times 480}$$
$$= 3.27\% \quad \text{OK}$$

Use 00 (1/0) wire for 480 V at 200 A.

Conduit Size

Three no. 00 (1/0) wires must be installed in a single conduit. What diameter should the conduit be?

From Table 6-1, line 00 (1/0), select the smallest conduit size that will carry three wires. The answer is 1½-in diameter conduit.

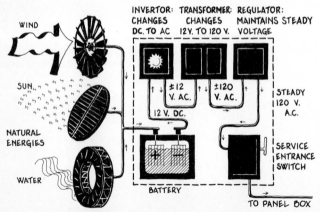

FIGURE 6-5 On-site generation of electricity.

ON-SITE GENERATION OF ELECTRICITY

Electricity may be created on the property by *wind* (mounting a generator onto a windmill), *water* (mounting a generator onto a waterwheel), or *sunshine* (photovoltaic cell or focusing collector). In each case, the generating equipment has a "nameplate" rating established by the manufacturer. (See Fig. 6-5.)

On-site generation of electricity is feasible only if desirable site conditions exist. These are described below.

Wind The generator and its revolving lever must be mounted on prominent terrain, such as a hilltop or ridge, and rise into constant breezes flowing across the property. The revolving lever should be mounted at least 15 ft above surrounding treetops and housetops, and the machinery should be able to turn and face winds blowing from any direction.

Water Swiftly flowing water must exist nearby. The liquid energy is dammed to create a vertical dimension of water weight or pressure which pushes downward to drive the generating equipment. This power may be undependable for small systems operating in regions having prolonged subfreezing winter weather.

Sunshine A high percentage of sunny weather must exist all year round. Also, no lofty terrain, surrounding foliage, or nearby architecture should cast shadows on the collector surfaces at any time of day from the sun's lowest trajectory across the sky in December to its highest trajectory in June.

As the creating of on-site generated wattage may be sporadic, the energy is emptied into storage batteries. Sizing these cells is an essential step in designing any system.

The electricity coming out of the batteries is direct current. This energy must be refined to be useful in the architecture. Transformers produce the desired voltage. Invertors change the energy from direct current to alternating current. Regulators maintain a steady voltage under varying load conditions. Switchgear and circuit breakers safeguard the system.

Below is the procedure for generating electricity on site.

1. Determine the average daily electrical load.
2. Determine the critical daily electrical load.
3. Determine the electric power potential of site.
   a. If wind
   b. If water
   c. If sunshine
4. Determine storage requirements of batteries.

Average Daily Electrical Load

The monthly electricity bills of a family of four average $67.43, and the electricity rate is 5.25 cents/kWh. What is the family's average daily electrical load?

$$L = \frac{E}{30R}$$

L = average daily electrical load in kilowatthours, ? kWh
E = amount of monthly electricity bill, $67.43
R = electricity rate in dollars per kilowatthour, 5.25 cents = $0.0525

$$L = \frac{67.43}{30 \times 0.0525}$$
$$= 42.8 \text{ kWh/day}$$

Critical Daily Electrical Load

What is the critical daily electrical load of the family living in the house of the problem above?

$$C = 2 + 1.5N$$

C = estimated critical daily electrical load (the amount of energy needed to operate refrigeration, hot water heating if electric, a few lamps, and essential electric appliances during prolonged emergency conditions, in kilowatthours), **?** kWh

N = number of people using the electricity, 4

$$C = 2 + 1.5 \times 4$$
$$= 8 \text{ kWh/day}$$

Wind power potential A family of four has analyzed their daily electric energy usage and found it to be 43 kWh average and 8 kWh critical. They believe that when necessary they can live comfortably on 20 kWh/day. Near their house is a low hilltop which would be a good place for a windmill. Tests reveal that 40 ft above the top of the hill the wind has an average speed of 12.5 mi/hr. Under these conditions, what size generator should their windmill have?

TABLE 6-2 DAILY POWER RATINGS FOR GENERATORS

Nom. output rating, W	Average windspeed, mi/hr					
	6	8	10	12	14	16
	Power rating, kWh/day					
500	0.3	0.7	1.0	1.3	1.6	1.8
1000	0.6	1.3	1.9	2.5	3.0	3.5
2000	1.2	2.3	3.5	4.6	5.8	6.8
4000	2.2	4.3	6.7	9.0	11.3	13.3
6000	3.3	6.7	10.1	13.6	17.1	20.6
8000	4.3	8.7	13.0	17.4	21.7	26.0
10,000	5.4	10.7	15.9	21.2	26.4	31.6
12,000	6.2	12.5	18.8	25.2	31.6	37.8

Analyze Table 6-2. Under 12.5 mi/hr (interpolate between 12 and 14 mi/hr), find the lowest wattage rating on the left that supplies at least 20 kWh/day. The answer is 10,000 W.

Water power potential   A family of five living in the foothills of western North Carolina has analyzed their daily electricity usage and found it to be 47 kWh average and 9.5 kWh critical. They want to install a water turbine on a stream flowing through a steep vale near their house. The owner, after examining the terrain around the stream, believes he can easily create a 46-ft head by building a small dam at the top of the vale, but the stream's rate of flow is only about 2 ft/sec. Under these conditions, how much electricity could the family draw from their stream?

$$P_w = 18K$$

P_w = water power potential in kilowatthours per day, ? kWh
K = electrical capacity of generator in kilowatts (from Fig. 6-6: at a head of 46 ft and a water flow rate of 2 ft/sec, the first diagonal line below this point), 4 kW

$$P_w = 18 \times 4$$
$$= 72 \text{ kWh/day}$$

This amount is much more than their daily needs require.

NOTE: Water turbines can generate great amounts of electricity from streams having high heads and low rates of flow.

Sunshine power potential   A lawyer in Durango, Colorado, owns a small house on several acres located deep in the San Juan mountains 50 mi away. Because the property has no electricity and it will cost several thousand dollars for the local utility company to bring it in from their nearest lines several miles away, the lawyer is considering installing photovoltaic cells on the roof of his cabin. He figures the cells

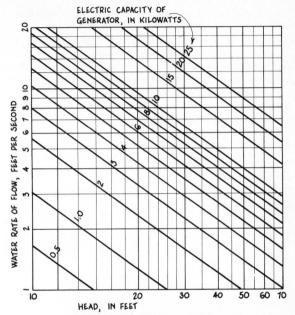

FIGURE 6-6 Water power graph.

could collect energy all week long and dump the wattage into storage batteries for his use from Friday night to Sunday night. The lawyer believes that two people would need only 5 kWh per weekend to lead a comfortable "Spartan" life at his mountain hideaway.

If the sun shines about 70 percent of the daylight hours there, what is the optimal surface area of the roof-mounted photovoltaic cells?

$$\text{Wattage used} = \text{wattage collected}$$
$$W_u = 0.0085FUCA$$

W_u = wattage used in the architecture (owner estimates this to be 5 kWh per weekend), $5 \times 1000 = 5000$ W

F = percentage of average sunshine for the region (as listed in local climate data; the site receives 70% of maximum sunshine), 70%

U = umbra fraction (that portion of collector surface shaded during the day; assume here that the panels are exposed to sunlight all day long), 1.0

C = capacity of photoelectric cells in watts per square foot (in 1985, available capacities run about 10 W/ft² per day), for a 1-week collection period, capacity = $10 \times 7 = 70$ W/ft²

A = area of photocell panels in square feet, ? ft²

$$5000 = 0.0085 \times 70 \times 1 \times 70 \times A$$
$$A = 120 \text{ ft}^2$$

Battery Storage

In the problem above, the lawyer's roof-mounted photovoltaic cells require storage batteries to hold the week's stored electricity for weekend use. If he uses 6-V golfcart batteries rated at 220 A each, how many should his system have?

$$S = 760CM \left(\frac{1}{N}\right)^{1.6}$$

S = total storage capacity of batteries in watts, ? W

C = critical daily electrical load (in this case, C is 5 kWh per weekend, 1 weekend = 2.33 days), $5/2.33 = 2.15$ kWh/day

M = maximum use factor (ratio of greatest to average daily use: Greatest daily use may be 3 kWh/day

"Average" daily use = $\dfrac{5 \text{ kWh per weekend}}{7 \text{ days per week}}$ = 0.72 kWh/day

$M = \dfrac{\text{Greatest daily use}}{\text{Average daily use}} = \dfrac{3.0}{0.72} = 4.17$

N = nongeneration factor (the portion of time the system may not generate electricity because of adverse climatic conditions; estimate this from knowledge of local climate; in most cases, typical values of N are as below):

> Wind, 0.75
>
> Water, 0.95
>
> Sunshine, use percentage of average sunshine/100 for the cloudiest month of the year, as listed in local climatic data

In this case, $N = 70/100 = 0.7$

$$S = 760 \times 2.15 \times 4.17 \times \left(\frac{1}{0.7}\right)^{1.6}$$
$$= 12,000 \text{ W}$$

How many batteries will be used?

$$\text{Wattage of 1 battery} = 6 \text{ V} \times 222 \text{ A} = 1320 \text{ W}$$
$$\text{Number of batteries} = \frac{12,000}{1320} = 9.09$$

Nine or ten batteries at 6 V and 220 A are needed.

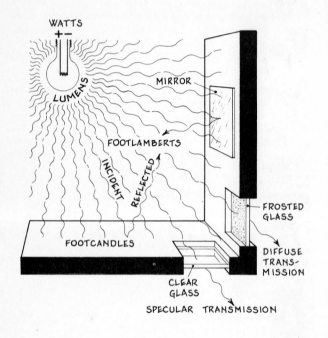

Illumination terms.

ILLUMINATION

GENERAL

Our ability to perceive interior space and perform work in it depends on the quantity of illumination reflecting off the various surfaces around us and entering our eyes. This visual strength relates to the amount of light falling on the surrounding surfaces, which depends on the quantity of light emanating from nearby fixtures, which relates to the wattage entering the lamps.

FUNDAMENTALS

Input, the amount of electrical energy entering a lamp, is measured in watts (W). *Output,* the amount of light energy leaving the lamp, is measured in lumens (lm). *Incidence,* the quantity of light falling on a nearby surface, is measured in footcandles (fc). *Reflectance,* the amount of light rebounding from that surface, is measured in footlamberts (fL). *Contrast,* the difference in amount of brightness between two adjacent surfaces, is measured as a ratio.

Input

How many amperes are used by a fluorescent fixture that contains four 40-W lamps and operates on 120 V?

$$W = AV$$

W = number of watts used by the fixture, $4 \times 40 = 160$ W
A = number of amperes used by the fixture, ? A
V = number of volts used by the fixture, 120 V

$$160 = A \times 120$$
$$A = 1.33 \text{ A}$$

Output

How many lumens are produced by a light fixture that contains four F48T12CW fluorescent light bulbs rated at 40 W each?

$$L = Nl$$

L = total output per fixture in lumens, ? lm
N = number of lamps or bulbs per fixture, 4
 l = illumination output per lamp (from Table 7-1, output of one 40-
       W F48T12CW lamp), 3000 lm

$$L = 4 \times 3000$$
$$= 12,000 \text{ lm}$$

TABLE 7-1 LAMP SPECIFICATIONS*

Type	Power, W	Output, lm	Efficacy, lm/W
Incandescent:			
A-19 60	60	855	14
A-21 100SB	100	1450	15
Quartz:			
T-H PAR-38	250	3220	13
T-H R-60	1000	17,000	17
Fluorescent:			
F48T12CW	40†	3000	75
F48T12CW/HO	60	4300	72
Mercury:			
R-40 H39BP-175/DX	175	5750	33
BT-46 H35NA-700	700	35,700	51
Sodium:			
BT-28	150	14400	96
T-18	1000	119,600	120

*These are only a few of the many lamps available. For further information examine manufacturers' product catalogs.

†These figures include energy consumed by the ballast.

Incidence

How many footcandles fall on a light green drafting surface whose average distance is 9 ft away from the center of a light fixture rated at 12,000 lm? (See Table 7-2.)

TABLE 7-2 LEVELS OF ILLUMINATION: ACTUAL OR REQUIRED

Light source or surface	Quantity of light, fc*
Sunlight, direct, near overhead, no haze	8400
Partly cloudy, cloud in front of sun	2700
Light overcast sky, sun near overhead	1200
Heavy overcast sky, late afternoon	500
Moonlight, full moon, clear sky	0.2
Natural light, 3 ft inside large window	100
Store:	
Window display area	200
Merchandising area	100
Office:	
General	100
Accounting and drafting	150
Conference room	50
Reading area	70
Dining area:	
Cafeteria	30
Intimate	10
Hospital:	
Patient's room	20
Operating room	2500
Church, worship area	15
Theatre:	
During performance	0.1
During intermission	5

TABLE 7-2 LEVELS OF ILLUMINATION: ACTUAL OR
REQUIRED *(Continued)*

Light source or surface	Quantity of light, fc*
Lobby, theatre, hotel	10
Recreation	30
Toilets, lockers, washrooms	30
Corridors, stairways, storerooms	20
Emergency exit lighting	1
Parking area	2

* All indoor measurements are taken 30 in above floor.

$$I = \frac{L}{D^2}$$

I = incident light upon a surface, ? fc
L = rated output of light fixture, 12,000 lm
D = distance between light source and surface, 9 ft

$$I = \frac{12000}{9^2}$$
$$= 148 \text{ fc}$$

Brightness

How many footlamberts of illumination emanate from a light green
drafting surface that has 148 fc incident upon it?

$$F = IC$$

F = brightness of light reflected from surface, ? fL
I = incident light upon the surface, 148 fc
C = coefficient of reflection of a surface (from Table 7-3, C of light
green surface), 0.63

$$F = 148 \times 0.63$$
$$= 93 \text{ fL}$$

TABLE 7-3 REFLECTION COEFFICIENTS OF SURFACES, C_r

Surface	C_r	Surface	C_r
Mirror	0.92	Aluminum	0.65
Glass:			
Clear	0.88	Wood	
Frosted	0.75	Light Oak	0.32
Plaster, white	0.91	Dark Oak	0.13
Paper, white	0.82	Mahogany	0.08
Acoustic tile, whitewhite	0.71	Cement, natural	0.25
Colors:		Brick, red	0.13
Colors:		Asphalt	0.07
White	0.81	Slate, dark grey	0.08
Cream	0.74	Snow:	
Light green	0.63	New	0.74
Light grey	0.58	Old	0.64
Tan	0.48	Earth:	
Dark grey	0.26	Moist cultivated	0.07
Olive green	0.17	Dry bare	0.40
Flat black	0.04	Vegetation, average	0.25
Stainless steel	0.60		

*These coefficients are representative values.

Contrast

What is the brightness ratio between a sheet of white illustration board and its surrounding light green drafting surface?

$$\text{Brightness ratio} = \frac{\text{higher } C_r}{\text{lower } C_r}$$

From Table 7-3:

    C_r of white paper, 0.82
    C_r of light green surface, 0.63

$$\text{Brightness ratio} = \frac{0.82}{0.63} = 1.30$$

NOTE: For visual comfort, the brightness ratio between two adjacent surfaces should not exceed 3.

Coefficient of Utilization

The coefficient of utilization, known as *CU*, is the portion of a room's lighting output that reaches the illumination surface, expressed as a value between 0 and 1. Actually there are two CUs: one for fixtures, and one for the room the fixtures are in. Manufacturers list fixture CUs in their product catalogs, and these values are used to find room CUs.

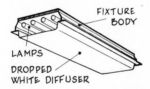

FIXTURE
BODY

LAMPS
DROPPED
WHITE DIFFUSER

FIGURE 7-1 Fluorescent lamp.

A room has fluorescent ceiling fixtures (Fig. 7-1), each of which contains four F48T12 CW lamps and has dropped white diffusers. The room is 21 by 40 ft in size, is 9 ft 4 in tall from floor to fixture face, has a white acoustic tile ceiling and cream-colored walls partially covered with dark oak bookcases, and contains 12 large drafting tables that have light green surfaces? What is the CU of the room?

$$CU = 0.9 \left(\frac{FRLW}{10H(L + W)} \right)^{0.2}$$

CU = coefficient of utilization of the room, **?**

  F = fixture CU (as listed in product catalog; a four-lamp fluorescent ceiling fixture with a white diffuser), 0.60

R = reflectance of room surfaces (from observation and Table 7-3, a rough estimate of the average reflectance of the room), about 0.6

L = length of room, 40 ft

W = width of room, 21 ft

H = height of room from floor to fixture face, 9 ft 4 in = 9.33 ft

$$\text{CU} = 0.9 \left(\frac{0.60 \times 0.6 \times 40 \times 21}{10 \times 9.33(40 + 21)} \right)^{0.2} = 0.9 \left(\frac{302}{5690} \right)^{0.2} = 0.50$$

NOTE: This method of calculating CU is only approximate, but in most circumstances it will be adequate. Absolute accuracy requires a lengthy analysis involving a specialist's expertise.

LIGHTING SYSTEMS

The formulas in this section are primarily for general spatial illumination. For task lighting and special effects, supplementary fixtures may be required. These may be designed according to the formulas given in the preceding section.

Number of Fixtures Required in an Area

An engineer's drafting room is 21 x 40 ft in size and has a CU of 0.50. To maintain a proper level of illumination, how many fluorescent fixtures containing four F48T12CW lamps each should be mounted in the ceiling? The floor-to-ceiling height = 9′ 4″.

$$IA = LNFMCR$$

I = required incident light upon the working surface (from Table 7-2), 150 fc

A = floor area of the space to be illuminated, $21 \times 40 = 840$ ft^2

L = illumination output per lamp (from Table 7-1, output of one F48T12CW lamp), 3000 lm

N = number of lamps per fixture, 4

F = number of fixtures required in the space, ? units

M = maintenance factor (varies according to conditions of cleanliness as described below):

> Enclosed clean fixture, 0.9
> Average conditions, 0.75
> Open dirty or dusty fixture, 0.6

Here we have enclosed fixtures; we'll assume average conditions of cleanliness for an $M = 0.75$

C = coefficient of utilization of room, 0.50

R = ratio of total room height to lamp-to-work surface height

Room height = 9 ft 4 in. Drafting desk work surfaces are typically 36 in above the floor.

Thus $R = 112/112 - 36 = 1.47$

$$150 \times 840 = 3000 \times 4 \times N \times 0.75 \times 0.50 \times 1.47$$
$$N = 19 \text{ fixtures}$$

NOTE: This formula may be used in noncontained spaces; for example, determining the number of fixtures in a single bay of a large area.

Area per Fixture

The lighting for the entrance lobby of a new hotel will be ceiling-mounted pendant diffusing spheres that contain one 250-W quartz PAR-38 lamp each (Fig. 7-2). If the lobby's CU is 0.31 and the fixture maintenance factor is 0.7, what is the ceiling area for each fixture?

$$IA = LNMC$$

I = required incident light upon the lobby area (from Table 7-2, I for a hotel lobby), 10 fc

A = floor area of space served by one fixture, ? ft^2

L = illumination output per lamp (from Table 7-1, L for a 250-W PAR-38 lamp), 3220 lm

N = number of lamps per fixture, 1 unit

M = maintenance factor, 0.7

C = coefficient of utilization for the space, 0.31

$$10 \times A = 3220 \times 0.7 \times 0.31$$
$$A = 69.9 \text{ ft}^2 \text{ per fixture}$$

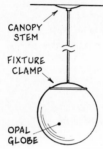

CANOPY STEM

FIXTURE CLAMP

OPAL GLOBE

FIGURE 7-2   Globe lamp.

Fixture Spacing

The spacing-to-mounting height ratio (S/MH) of a certain ceiling fixture is listed in its catalog as 1.5. If these lights are mounted in a ceiling 9.5 ft above the floor, how far apart should they be?

$$S/MH = \frac{S}{H}$$

S/MH = spacing-to-mounting height ratio, 1.5 units

S = side-to-side spacing of fixture, ? ft

H = height of fixtures above floor, 9.5 ft

$$1.5 = \frac{S}{9.5}$$
$$S = 14.3 \text{ feet apart}$$

EXTERIOR ILLUMINATION

Exterior lighting systems have been developed to enable outdoor environments to be seen at night. Their flow of brightness usually reveals a space differently than sunlight: sunlight casts shadows downward, whereas night lighting generally casts shadows upward. For best results, the principle lighting of a surface should come from one main direction that forms an angle of at least 30° to the surface; the lighting should be located at least 15° from the observers' principal line of view. Floodlighting is most effective if the illuminated surface is light in color, rough in texture, and has a dark background. No night light source should be visible to the observer. (See Table 7-4.)

TABLE 7-4 ILLUMINATION LEVELS FOR FLOODLIGHTING, fc

Surface material	Bright surroundings	Dark surroundings
Light marble, white or cream terra cotta, white paint or plaster	15	5
Concrete, tinted stucco, light grey and buff-colored surfaces	20	10
Common tan brick, sandstone, medium grey surfaces	30	15
Common red brick, brownstone, dark colors, stained wood shingles	50	25

Night Lighting

The elders of a historical landmark church near the center of a rural village wish to make the tall steeple of their place of worship a nighttime focal point by floodlighting it. Luminaires will be mounted near the peaks of the four roofs radiating out from the steeple's base and shine on the steeple's four sides, which are built of wood painted white (Fig. 7-3). If the luminaires will form a 60° angle between the steeple's cornice and its base, which are 42 vertical ft apart, how powerful should the luminaires be?

FIGURE 7-3 Church at night.

STEP 1. Find the distance between the luminaire and the steeple's cornice, as this is the maximum distance between the fixture and the part of the facade to be flooded (see Fig. 7-4).

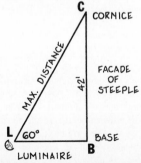

FIGURE 7-4 Church steeple dimensions.

$$\sin 60° = \frac{BC}{CL} = \frac{42}{CL} = 0.866$$

$$CL = \frac{42}{0.866} = 48.5 \text{ ft}$$

STEP 2. From Table 7-4, find the recommended floodlighting incident upon the facades of the steeple. From Table 7-4, recommended floodlighting incident upon white painted surface with dark surroundings is 5 fc.

STEP 3. Find the output of each luminaire from the following formula.

$$I = \frac{L}{D^2}$$

I = incident light upon a surface, 5 fc
L = rated output of light fixture, ? lm
D = distance between light source and surface, 48.5 ft

$$5 = \frac{L}{48.5^2}$$

$$L = 5 \times 48.5^2 = 11,800 \text{ lm}$$

The minimum output per luminaire is 11,800 lm.

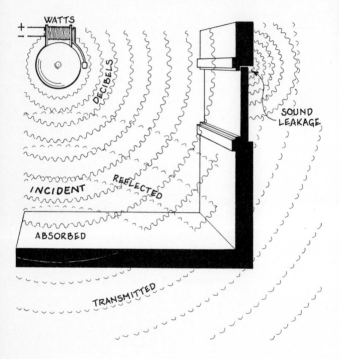

Acoustic terms.

ACOUSTICS

GENERAL

Sound is a vibration in an elastic medium. These oscillations—whether fast, slow, hard, or soft—are always subtle and often unpredictable. Sound waves can turn the environment they travel through into an acoustic landscape that can be as ecstatic or unbearable as any visual landscape. Indeed, an otherwise beautiful architectural space can be made uninhabitable by a seemingly trivial oversight in acoustic design.

BASICS

Sound has three primary properties: velocity, pitch, and power. *Velocity* is the speed of sound waves measured in feet per second. This speed varies greatly according to the medium the waves travel in and slightly according to the temperature of the medium. *Pitch* is the vibration frequency of a sound measured in cycles per second, or hertz (Hz). The frequency may be fast, as in the shrill blow of a whistle, or slow, as in the sound of a bass voice. *Power* is the loudness of a sound measured in decibels (dB). This quality diminishes as the sound travels outward from its source.

Velocity
What is the speed of sound at 68° F?

$$V = 1087 \sqrt{1 + \frac{F - 32}{459}}$$

V = velocity of sound, ? ft/sec
F = temperature of air in degrees Fahrenheit, 68° F

$$V = 1087 \sqrt{1 + \frac{68 - 32}{459}}$$
$$= 1130 \text{ ft/sec}$$

Frequency

What are the lengths of two sound waves traveling at room temperature, one at 125 Hz, the other at 20,000 Hz?

$$W = \frac{V}{F}$$

W = length of sound wave, ? ft
V = velocity of sound wave (in this case, at room temperature, 68°F), 1130 ft/sec
F = frequency of sound wave:
  F_1 = 125 Hz
  F_2 = 20,000 Hz

125 Hz:

$$W = \frac{1130}{125} = 9.04 \text{ ft}$$

20,000 Hz:

$$W = \frac{1130}{20000} = 0.0565 \text{ ft, or } 0.678 \text{ in}$$

Figure 8-1 presents a spectrum of sound frequency levels.

NOTE: Certain architectural materials absorb sound better at one frequency than at others. Most hard materials, such as glass and concrete, absorb low sounds better because the sounds' longer waves have more momentum to penetrate dense masses. Thus, low-pitched sounds pass through walls and floors better than do high sounds. On the other hand, soft materials, like carpet and drapery fabric, are best at absorbing high sounds because the light short waves are more apt to get caught in the "nap" of the fabrics' rough, pliable surfaces. These short waves though, because of their low momentum, usually cannot penetrate deeply into a material's mass and are easily dissipated.

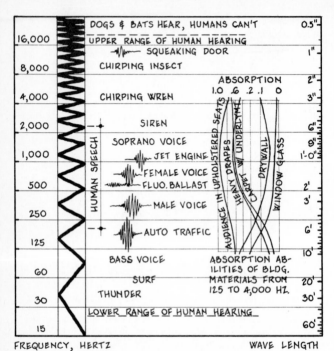

FIGURE 8-1   Sound frequency levels.

Power
See Fig. 8-2.

   Relation between sound and distance   A loudspeaker has a power of 110 dB at its source. How powerful is the sound 50 ft away?

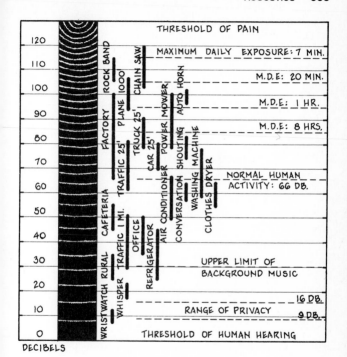

FIGURE 8-2 Sound power levels.

$$P = S - 0.7 - 20 \log D$$

P = power of sound at a distance from its source, ? dB
S = power of sound at source, 110 dB
D = distance between source and receiver, 50 ft

$$P = 110 - 0.7 - 20 \log 50$$
$$= 110 - 0.7 - 34.0$$
$$= 75.3 \text{ dB 50 ft from source}$$

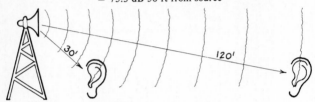

FIGURE 8-3 Airhorn on tower.

Relative sound power An air horn on a tower measures 102 dB 30 ft away (Fig. 8-3). How powerful is the sound 120 ft away?

$$P_1 - P_2 = 20 \log \frac{D_2}{D_1}$$

P_1 = near sound power, 102 dB
P_2 = far sound power, ? dB
D_1 = near distance from source, 30 ft
D_2 = far distance from source, 120 ft

$$102 - P_2 = 20 \log \frac{120}{30}$$
$$= 12.0$$
$$P_2 = 102 - 12 = 90 \text{ dB}$$

NOTE: Doubling sound distance reduces power by approximately 6 dB. Halving sound distance increases power by approximately 6 dB.

Adding sound powers A 90-dB loudspeaker mounted on a boom for an outdoor concert is not considered loud enough. A 100-dB and a 110-dB speaker are added to the boom. What is the total power of the three speakers?

$$P_t = P_s + 10 \log \frac{S}{P_s}$$

P_t = total power of all speakers, **?** dB
P_s = power of strongest speaker, 110 dB
S = sum of the power levels of all the speakers, $90 + 100 + 110 = 300$

$$P_t = 110 + 10 \log \frac{300}{110}$$
$$= 110 + 4.4 = 114 \text{ dB}$$

NOTE: Doubling the power of a sound increases its loudness by approximately 3 dB. Halving its power decreases its loudness by approximately 3 dB.

Relation between sound and electricity   How powerful is the sound 35 ft away from a loudspeaker rated at 20 W and operating at full volume?

$$P_r + 20 \log D = 10 \log W + 119$$

P_r = power of sound at receiver, **?** dB
D = distance between source and receiver, 35 ft
W = wattage of speaker, 20 W

$$P_r + 20 \log 35 = 10 \log 20 + 119$$
$$P_r + 30.9 = 13.0 + 119$$
$$P_r = 101 \text{ dB}$$

INCIDENCE

When a sound strikes a surface, it is either absorbed or reflected.

The sound absorbed by a material is measured in sabins. Not only do different materials have different sound absorption rates, the rates for each material often vary according to the sound's frequency.

Of the sound absorbed by a material, a portion passes through its mass into adjoining spaces. The difference between the entering sound and the exiting sound is known as the material's *STC rating* and is measured in decibels. For example, a wall that receives a sound on one side at 60 dB and transmits it into the space on the other side at 16 dB has an STC rating of 44 dB (Fig. 8-4).

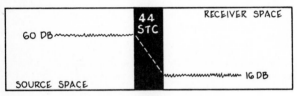

FIGURE 8-4 Sound through wall.

Every kind of floor, wall, and ceiling construction can be given an STC rating. Usually, the higher the STC, the more expensive is the construction.

The portion of sound reflected by a surface enters the space it came from and continues to carom off the surrounding surfaces until the sound has dissipated. The amount of time this degeneration takes is known as *reverberation time* and is measured in seconds. The more reflective the surfaces of a space, the higher its reverberation time.

Absorption

How much sound is absorbed in a conference room 32 ft long, 20 ft wide, and 10 ft high? The floor is concrete, one long wall is window glass, and the other three walls are gypsum board nailed to 2 x 4 studs; the ceiling is smooth plaster.

$$\alpha_t = \alpha A_1 + \alpha A_2 + \alpha A_3 + \cdots \alpha A_n$$

α_t = total absorption of sound by the space, **?** sabins

α = Coefficient of sound absorption for each material; select from Table 8-1; because the sound generating in the space will be primarily human conversation, which from Fig. 8-1 occurs at about 500 Hz, use the absorption coefficients for 500 Hz

TABLE 8-1 ABSORPTION COEFFICIENTS OF MATERIALS per ft^2*

Material	125 Hz	500 Hz	4000 Hz
Brick:			
Unglazed	0.03	0.03	0.07
Unglazed, painted	0.01	0.02	0.03
Carpet:			
Heavy on concrete	0.02	0.14	0.65
With rubber underlayment	0.08	0.39	0.63
Concrete block:			
Coarse	0.36	0.31	0.25
Painted	0.10	0.06	0.08
Drapes:			
Light velour	0.03	0.11	0.35
Heavy velour	0.14	0.55	0.65
Concrete or terrazzo floor	0.01	0.015	0.02
Asphalt tile on concrete	0.02	0.03	0.02
Wood	0.15	0.10	0.07
Glass:			
Plate	0.18	0.04	0.02
Window	0.35	0.18	0.04
Gypsum, 0.5 in on 2 x 4 studs	0.29	0.05	0.09
Marble or glazed tile	0.01	0.01	0.02
Plaster:			
Smooth on brick	0.13	0.02	0.05
Rough or smooth on lath	0.02	0.03	0.03

TABLE 8-1   *(Continued)*

Material	125 Hz	500 Hz	4000 Hz
Plywood panelling, ⅜ in	0.28	0.17	0.11
Acoustical tile on suspended ceiling	0.40	0.50	0.60
Audience in upholstered seats	0.60	0.88	0.85
Empty upholstered seats	0.49	0.80	0.70
Empty metal or wood seats	0.15	0.22	0.30

*These numbers may vary slightly depending upon a material's condition, age, and manufacturing process.

A = surface area of each material enclosing the space

Below is a list of the materials enclosing the space:

Surface	*α at 500 Hz*	*Area, ft^2*	
Concrete floor	0.015	$32 \times 20 = 640$	
Window glass	0.18	$32 \times 10 \times 0.5 = 160$	
Gypsum wall	0.05	$32 \times 10 \times 0.5 = 160$	⎫
		$32 \times 10 = 320$	⎬ 880
		$20 \times 10 \times 2 = 400$	⎭
Plaster ceiling	0.02	$32 \times 20 = 640$	

$$\alpha_t = 0.015 \times 640 + 0.18 \times 160 + 0.05 \times 880 + 0.02 \times 640$$
$$= 9.6 + 28.8 + 44 + 12.8 = 95.2 \text{ sabins}$$

Reverberation

What is the reverberation time for the 32 x 20 x 10 ft conference room described in the previous problem? Is this a satisfactory time for this space? If not, what remedies should be taken?

$$R = \frac{0.05V}{\alpha_t}$$

R = reverberation time, ? sec
V = volume of the space, $32 \times 20 \times 10 = 6400$ ft^3

α_t = total absorption of enclosing surfaces (from previous problem), 95.2 sabins

$$R = \frac{0.05 \times 6400}{95.2}$$
$$= 3.36 \text{ sec}$$

Is this a satisfactory reverberation time for this space? Consult Table 8-2. From Table 8-2, we can see that 3.36 sec is *not* a desirable reverberation time for a conference room.

TABLE 8-2 DESIRABLE REVERBERATION TIMES FOR SPACES

Space	Time, sec
Offices, homes, small private spaces	0.3–0.5
Library reading rooms	0.5–0.7
Broadcast studios for speech	0.4–0.6
Elementary classrooms	0.6–0.8
Lecture and conference rooms	0.9–1.1
Auditoriums	1.5–1.8
Community church naves	1.4–2.0
Concert halls	1.4–1.6
Opera halls	1.5–1.8
Cathedral nave	2.8–3.4
Movie theatre	0.8–1.2

What remedies should be taken? The room's surfaces are too reflective for its purpose. Try installing carpet with rubber underlayment on the concrete floor and hanging floor-to-ceiling light drapes against the long wall without glass. Now, first recalculate the sound absorption of the enclosing surfaces, then recalculate the reverberation time.

$$\alpha_t = \alpha A_1 + \alpha A_2 + \alpha A_3 + \cdots \alpha A_n$$

Below is a list of the materials enclosing the space:

Surface	At 500 Hz	Area, ft²
Carpet floor	0.39	$32 \times 20 = 640$
Window area	0.18	$32 \times 10 \times 0.5 = 160$
Gypsum wall	0.05	$32 \times 10 \times 0.5 = 160$
		$20 \times 10 \times 2 \quad = 400$ } 560
Drapery wall	0.11	$32 \times 10 = 320$
Plaster ceiling	0.02	$32 \times 20 = 640$

$$\alpha_t = 0.39 \times 640 + 0.18 \times 160 + 0.05 \times 560$$
$$+ 0.11 \times 320 + 0.02 \times 640$$
$$= 250 + 28.8 + 28 + 35.2 + 12.8 = 355 \text{ sabins}$$
$$R = \frac{0.05V}{\alpha_t} = \frac{0.05 \times 6400}{355}$$
$$= 0.90 \text{ sec} \quad \text{OK}$$

Transmission

Example 1   It is desired that the sound passing through a wall having an acoustic intensity level of 65 dB on one side not be above 16 dB (considered the minimum power level to ensure privacy) as it enters the space on the other side. What should be the minimum STC rating of the intervening wall?

$$\text{STC} = P_s - P_r$$

STC = rating of barrier construction, **?** dB
  P_s = acoustic power level of source space, 65 dB
  P_r = acoustic power level of receiving space, 16 dB

$$\text{STC} = 65 - 16$$
$$= 49 \text{ dB}$$

Example 2 A floor construction has a standard carpet on rubber underlayment on a plywood subfloor on 1 x 3 sleepers on a ½-in fiberboard resilient mat on 3-in concrete on cellular metal decking. What is the STC rating of the floor if 16 in below the underside of the cellular metal there is a suspended ceiling on wire hangers?

$$STC = S_b + S_1 + 0.5(S_2 + S_3 + S_4 + \cdots S_n)$$

STC = STC rating of total construction, ? dB
S_b = STC rating of basic construction (from Table 8-3, STC of basic construction for concrete on cellular metal decking), 38 dB

TABLE 8-3 STC AND IIC RATINGS FOR CONSTRUCTION TYPES, in dB*

Material	STC	IIC
STUD WALL CONSTRUCTION		
¼-in plywood both sides, on 2 x 4 studs	24	
½-in drywall both sides:		
on 2 x 4 wood studs	32	
on 12-in steel channel chase partition	42	
⅝-in drywall both sides:		
on 2 x 4 wood	34	
on 2 x 4 wood staggered on 2 x 6 plates	42	
on double 2 x 4 wood, 2-in between partitions	43	
on 2½-in steel studs	37	
on 3½-in steel studs	39	
ADD:		
3½-in batts to voids	+ 5	
9-in batts to deep voids	9	
Double drywall:		
one side	6	
both sides	9	

TABLE 8-3 (Continued)

Material	STC	IIC
Resilient channels:		
one side	6	
both sides	9	
½-in fiberboard:		
one side	4	
both sides	6	
MASONRY WALL CONSTRUCTION		
6-in lightweight concrete block, plain	33	
8-in standard concrete block, plain	50	
ADD:		
½-in drywall on 1 x 2 furs:		
one side	+ 7	
both sides	10	
½-in drywall on resilient channels:		
one side	12	
both sides	15	
1½-in batts in furring cavity:		
one side	+ 3	
both sides	5	
Fill cores with sand	3	
Paint or plaster:		
one side	2	
both sides	4	
OTHER MATERIALS		
Aluminum, 0.025 in thick	19	
Sheet steel, 18 gauge	30	
Lead, 0.0625 in thick	34	
Plate glass, ¼ in thick	26	
Insulating glass, ¼ in glass ½ in air, ¼ in glass	32	

TABLE 8-3 (Continued)

Material	STC	IIC
Glass block, 3¾ in thick	40	
Door, wood, 1¾ in thick:		
hollow core	26	
solid core	29	
Door, metal, 1¾ in thick:		
hollow	30	
packed	32	
FLOOR CONSTRUCTION:		
Plywood subfloor on wood joists with drywall under	41	36
4-in reinforced concrete slab	41	33
6-in reinforced concrete slab	46	35
Concrete pad on cellular metal decking	38	29
ADD:		
Vinyl tile or linoleum finished floor	2	2
Hardwood or parquet floor	2	3
Lightweight wall-to-wall carpet on pad	4	10
Standard wall-to-wall carpet on pad	5	16
1½-in concrete slab	7	3
½-in fiberboard between finish and rough floors	3	3
Wood floor:		
on 1 x 3 sleepers	8	6
on fiberglass insulation	8	10
on ½-in fiberboard	10	7
Fiberglass insulation in voids of wood framing	2	2
Drywall to resilient channels on underside	5	8
Suspended ceiling on resilient clips or wire hangers, 10-in		
min. airspace	12	8
Plaster or paint on underside of concrete	2	0

*With these ratings, variations of 2 to 3 dB are considered slight, 5 to 6 dB are noticeable, and 10 to 15 dB are dramatic.

S_1 = STC rating of additional construction having the highest STC; to find this, make a list of all additional constructions and their STC ratings:

Added construction	STC
1. Standard carpet on underlayment	5
2. Wood subfloor on 1 x 3 sleepers on fiberboard	10
3. Suspended ceiling on hangers	12

S_1 = 12 dB

$S_2, S_3, S_4, \ldots S_n$ = STC ratings of all additional constructions, S_2 = 10 dB, S_3 = 5 dB

$$\text{STC} = 38 + 12 + 0.5(10 + 5)$$
$$= 57.5$$

Example 3 The assembly line area on the floor directly under a factory manager's office has an acoustic intensity level of 81 dB. If the manager's office is 12 x 19 ft in size, its enclosing surfaces absorb 116 sabins, and its desired privacy level is 13 dB, what is the minimum STC rating of the floor and ceiling construction between the office and assembly area?

$$P_s = \text{STC} + 10 \log \frac{A}{\alpha_t} + P_r$$

P_s = acoustic power level of source space, 81 dB

STC = rating of barrier construction, ? dB

A = surface area of barrier construction (floor area of room), 12×19 = 228 ft^2

α_t = total sound absorption of enclosing surfaces of receiving space, 116 sabins

P_r = acoustic power level of receiving space (level of privacy desired), 13 dB

$$81 = STC + 10 \log \frac{228}{116} + 13$$
$$= STC + 2.9 + 10$$
$$STC = 81 - 2.9 - 10 = 65.1 \text{ dB minimum}$$

Masking

The factory owner of the previous problem has decided that the wall rated at STC = 65 dB will be too expensive to build. He now wants intercom music introduced to mask part of the sound emanating from the assembly lines on the floor below. If the background music has a masking level of 30 dB, what is the new STC rating for the floor and ceiling construction between the office and assembly area below? Remember, the office is 12 x 19 ft in size and its enclosing surfaces absorb 116 sabins.

$$P_s = STC + 10 \log \frac{A}{\alpha_t} + P_r + M - 10$$

P_s = acoustic power level of source space, 81 dB

STC = rating of barrier construction, ? dB

A = surface area of barrier construction (floor area of room), 12 × 19 = 228 ft^2

α_t = total sound absorption of enclosing surfaces of receiving space, 116 sabins

P_r = acoustic power level of receiving space (level of privacy desired), 13 dB

M = masking level of background sound, 30 dB

$$81 = STC + 10 \log \frac{228}{116} + 13 + 30 - 10$$
$$STC = 81 - 2.9 - 13 - 30 + 10 = 45.1 \text{ dB}$$

See Fig. 8-5 for the effects of the masking.

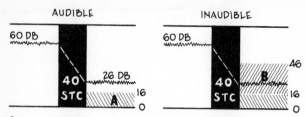

A. O TO 16 DB = RANGE OF PRIVACY: NOT ENOUGH TO MASK 26 DB SOUND ENTERING SPACE.

B. 16 TO 46 DB = RANGE OF MASKING: ADDED TO RANGE OF PRIVACY, IT HIDES 26 DB SOUND ENTERING SPACE.

FIGURE 8-5   Effects of masking undesirable sound transmittance.

SOUND LEAKAGE

Like a chain, good acoustic design is only as strong as its "weakest link." A tiny opening, such as an uncaulked seam between a duct and the wall it passes through, the grill of a heating unit located above a suspended ceiling, or two electrical outlets mounted back to back in a wall, can allow more sound to pass through than is dissipated by much good design and construction. In fact, the sound traveling through 1 in^2 of open area will equal the total noise reduced by 25 ft^2 of an STC 46 rated wall.

Sound Transmission through Openings

A door that has loosely fitting trim and a ⅜-in gap at its bottom is located in an 8 x 15 ft wall rated at STC 48. Considering the sound escaping around the door, what is the "new" STC rating of the wall?

STEP 1.   Estimate the open area around the door caused by the loose trim and gap at the bottom.

Top and sides:

$$\text{Space between door \& trim} = \pm \tfrac{1}{16} \text{ in wide}$$
$$\text{Length of trim} = 80 + 30 + 80 = 190 \text{ in}$$
$$\text{Open area} = 1/16 \times 190 = 11.9 \text{ in}^2$$

Bottom:

$$\text{Gap of open area} = 3/8 \times 30 = 11.2 \text{ in}^2$$

STEP 2.   Find the percentage of open area in wall.

$$P = \frac{100 \times \text{area of openings}}{\text{total area of wall}} = \frac{100(11.9 + 11.2)}{8 \times 15 \times 144}$$
$$= 0.13\%$$

STEP 3.   From Fig. 8-6, find the decibels to be subtracted from the wall STC rating. The effective transmission loss at a barrier STC of 48 and open area percentage of 0.13 equals 18 dB. Thus, effective STC of the wall is

$$48 - 19 = 30 \text{ dB}$$

Impact Noise

An office building will have a woman's modeling agency as a new tenant on one floor. The building's manager knows from past experience that women walking in high heels can create a disturbing impact noise that penetrates the 4-in thick concrete floor with suspended ceiling and enters spaces below. As this manager's realty offices will be located directly underneath the new agency, he wants to make sure their renovation work will eliminate any noises that could disturb his activities and those of his staff. In this case, what recommendations should the architect make?

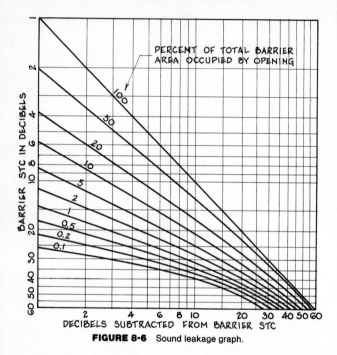

FIGURE 8-6 Sound leakage graph.

STEP 1.   Calculate the impact isolation class (IIC) of the construction from the formula below.

$$I_s = \text{IIC} + I_r + M - 10$$

I_s = impact noise level at source (from Table 8-4, I_s for spiked high heeled shoes), 74 dB

IIC = impact isolation class of the construction, ? dB

I_r = impact level desired (ranges from a silent 1 dB to a gently audible 16 dB); in this case, it would be wise to use 1 dB

M = masking effect of background sound level; recommend installing in realty office intercom music system whose masking effect is rated at a conservative 25 dB

$$74 = IIC + 1 + 25 - 10$$
$$IIC = 58 \text{ dB}$$

TABLE 8-4 IMPACT NOISE LEVELS*

Activity	Typical impact loudness, dB
Water hammer	94
Heavy cart rolled across floor	71
Table shoved into wall	73
Foot tapping to music	64
200-pound adult jumping	77
Chair scraping across hardwood or concrete floor	68
Small dropped object	67
Large dropped object	73
Man's footsteps:	
hard leather heels	63
rubber heels	58
Woman's footsteps:	
spiked heels	74
2-in high broad heels	60

*Representative values

STEP 2. Select a type of construction from Table 8-3 that satisfies the above IIC condition in the formula below.

$$IIC = I_b + I_1 + 0.5(I_2 + I_3 + I_4 + \cdots + I_n)$$

IIC = impact isolation class of floor construction from step 1, 58 dB minimum

I_b = IIC rating of basic construction (from Table 8-3, IIC of 4-in concrete slab), 33 dB

I_1 = IIC rating of additional layer of construction having the highest IIC (For this make a list of all additional constructions and their IIC ratings from Table 8-3. This includes existing constructions and whatever future constructions are necessary to give the final architecture an IIC rating of at least 53 dB. This must necessarily be done by trial and error.)

Added construction   IIC

1. Suspended ceiling on wire hangers   8
2. Standard wall-to-wall carpet on pad   16
3. Wood floor on 1 x 3 sleepers on fiberglass insulation   10

$$I_1 = 16 \text{ dB}$$

$I_2, I_3, I_4, \ldots I_n$ = IIC ratings of all additional constructions listed above: $I_2 = 10$ dB, $I_3 = 8$ dB

$$58 \leq 33 + 16 + 0.5(10 + 8) \leq 58 \qquad \text{OK}$$

Architect's recommendation Install in the modeling agency a floor of wall-to-wall carpet on wood subfloor on 1 x 3 sleepers on fiberglass insulation; install an intercom music system in the ceiling of the realty office.

Other Leakage Considerations

Several other types of sound leakage occur. One example is flanking paths (sound passing around a barrier between source and receiver, such as over a wall that ends above a suspended ceiling between two rooms). Another example is structure-borne sound (sound travelling through the solid parts of a building, such as vibrations from rigidly mounted mechanical equipment which pass through the building's structure into adjoining spaces). For the most part, these problems cannot be quantified mathematically, but the following suggestions will help one develop "sound" judgement in dealing with them.

1. On floor plans, arrange noisy spaces in one area and quiet spaces in another; locate minor spaces such as corridors and closets between them as buffers.
2. In section, arrange noisy functions on top of each other and do the same with quiet functions. For example, in apartments stack kitchens over kitchens, baths over baths, and bedrooms over bedrooms.
3. In acoustically important areas avoid parallel and concave surfaces (whether wall to wall or floor to ceiling), long narrow spaces, near-cubical volumes, and low-ceilinged rooms of large floor area. Instead, use convex surfaces, "lightning bolt" wall plan shapes, angled walls, or sloping ceilings.
4. Choose quietly operating equipment and components. Air conditioners should have quiet motors and balanced fans. Water pipes and valves should be adequately sized to minimize whistling. Fluorescent lights should have quiet ballasts.
5. Locate mechanical equipment far from study or sleeping areas and isolate the machinery in a room that has a gasketed solid core door. Also, anywhere that noise-making equipment or extension thereof touches the surrounding architecture, break *every* contact with some sort of acoustic padding such as resilient mounts, duct bellows, tube expansion valves or flexible loops, resilient seals, flexible packing, and the like.
6. Where hallways have doors on both sides, do not locate the openings directly across from each other. Where noise control is critical, avoid sliding doors. Instead, use solid wood core doors with seals at top, soft weather-stripping on the sides, and automatic threshold closers on the bottom.
7. Noise transmission through windows can be reduced with thick glass and double glazing. Openable windows should close tightly

and be weatherstripped. Locate windows so that sound cannot travel directly out one opening and into another, and minimize glass areas where they face noisy activities.

OUTDOOR SOUND BARRIERS

A four-lane expressway passes 200 ft behind the back of a residence. The owner wants to install a barrier across the back of her property that will reduce the noise reaching her house. In this case, what recommendations should the architect make?

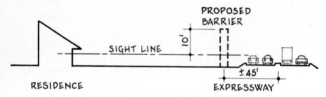

FIGURE 8-7   Outdoor sound barrier.

Initial recommendations: A good outdoor sound barrier (Fig. 8-7) rises considerably above the line of sight between source and receiver and is located close to either one. The barrier's effectiveness may be estimated by the following formula.

$$NR = 10 \log \frac{H^2}{D} + 10 \log F - 17$$

NR = noise reduction due to barrier, ? dB

  H = height of barrier; count only the height above the sight line between source and receiver (about 5 ft above ground); a 15-ft high barrier (H = 10 ft) would be about as high as could economically be built

D = distance from barrier to source or barrier to receiver (use whichever is less, in this case barrier to source: measure to center of
source of sound, the middle of expressway), about 45 ft

F = frequency of the sound (from Table 8-1, general frequency of
vehicular traffic), about 250 Hz

$$NR = 10 \log \frac{10^2}{45} + 10 \log 250 - 17$$
$$= 3.5 + 24.0 - 17 = 10.5 \text{ dB}$$

NOTE: Such noise reduction increases 3 dB for every octave the frequency increases. Thus, at a frequency of 500 Hz the noise reduction
would be 13.5 dB, and at 1000 Hz it would be 16.5 dB. Thus, a barrier
of this kind reduces sound more effectively at higher frequencies.

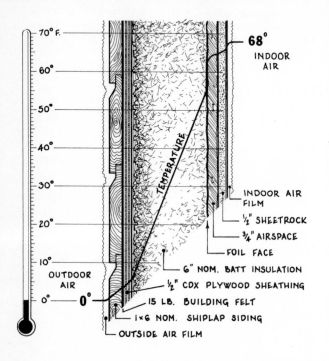

70° F.

60°

50°

40°

30°

68°
INDOOR
AIR

TEMPERATURE

20°

10°

OUTDOOR
AIR

0°   0°

INDOOR AIR
FILM

½" SHEETROCK

¾" AIRSPACE

FOIL FACE

6" NOM. BATT INSULATION

½" CDX PLYWOOD SHEATHING

15 LB. BUILDING FELT

1×6 NOM. SHIPLAP SIDING

OUTSIDE AIR FILM

Wall section.

HEAT FLOW

GENERAL

A comfortable indoor environment has a year-round temperature between about 67 and 77°F. When outdoor temperatures range higher or lower, heat flows from the warmer area to the colder. The greater the temperature difference between the two areas, the faster the heat flow. The thicker the insulation between the two areas, the slower the heat flow. Quantifying these dynamics is essential for sizing the architecture's heating and cooling systems. The amount of heat flowing out through the building envelope at the winter design temperature determines the size of the architecture's heating system, while the amount of heat flowing in through the building envelope at the summer design temperature determines the size of the cooling system.

These mathematics involve many unknown factors that are inextricably intertwined. Because of this, the problems in this chapter are not divided into small individual parts as are those of other chapters but instead exist as part of a lengthy sequence that must be followed from beginning to end before a building's heating or cooling load is adequately defined. This analysis requires calculating the design heating and cooling loads of every type of construction in every envelope surface of every perimeter space in the architecture. In large buildings, this may involve hundreds of calculations.

A good way to organize these mathematics is to make a foldout drawing of the building envelope, then label the different rooms and surfaces in each. For example, if the architecture is a residence (Fig. 9-1), one might do the following.

1. Remove all unheated areas from the architectural shape.
2. Open up the envelope into a flat plane, then label all rooms and construction types.

If the architecture is an office building (Fig. 9-2), a foldout of its envelope might be prepared as shown on page 360.

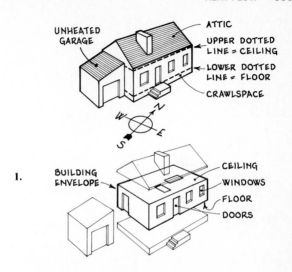

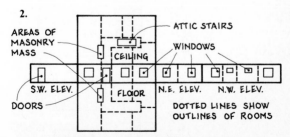

FIGURE 9-1 Residential building envelope.

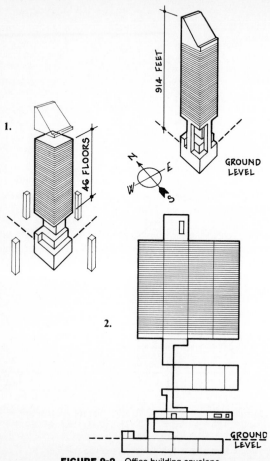

FIGURE 9-2 Office building envelope.

1. Remove all unheated areas.
2. Open up the envelope, then label rooms and construction types.

Heat flow occurs in five ways:

1. Conduction
2. Infiltration
3. Auxiliary
4. Pickup heating load
5. Solar heat gain

The following sections describe each of these ways in greater detail.

CONDUCTION

Conduction is heat flow through all solid parts of the building envelope. This includes insulated and noninsulated surfaces, glass, and doors. Conduction heat flow is described in the following formulas.

Cold weather:

$$C_c = \frac{A}{RP}(T_i - T_o)t$$

Hot weather:

$$C_h = \frac{AF}{RP}(T_o - T_i)t$$

C_c = total conduction heat loss through one type of construction type in the building envelope during cold weather, in British thermal units (BTU) per hour

C_h = total conduction heat loss through one type of construction in the building envelope during hot weather, in BTU per hour

A = surface area of each construction type in the building envelope in square feet (for warm-weather calculations, do not consider heat gains through parts of the building envelope that exist more than one foot below ground level).

R = thermal resistance, or insulating strength, of each construction type in the building envelope; usually equals the R value of the insulation plus about 3 (R values of various insulations, plus those of material assemblies not conforming to this rule, are listed in Table 9-1). $R = 1/U;$ $U = kt.$

TABLE 9-1 THERMAL RESISTANCE OF BUILDING MATERIALS

Insulation	Resistance
Fiberglass batts	3.5 per inch T
Styrofoam boards	4.2
Urethane boards	6.2
Isocyanurate boards with reflective surface*	8.0
Acoustical tile	2.7
Corkboard	3.5
Mineral fiberboard	2.8
Mineral wool loose fill	3.3
Perlite loose fill	2.6
Vermiculite loose fill	2.1
Reflective surface in noninsulated void*	1.0
Glass:	
Single pane	1.7 per unit†
Thermopane	2.5
Triplepane	3.3
Storm windows and single pane glass	1.8

*For reflective surfaces to be effective they must be exposed to a ½ to ¾-in thick still airspace.

†These values include inside and outside airfilms

P = perimeter heat flow fraction (in almost any insulated construction, part of the surface area is not insulated; this requires an adjustment factor, as listed for various types of construction in Table 9-2)

F = Fan factor, if a ventilation fan exists within the envelope materials (such as between a roof and suspended ceiling, $F = 0.75$; otherwise $F = 1$.

T_i = indoor temperature (typical indoor temperatures are 68°F in cold weather and 78°F in hot weather)

T_o = outdoor temperature in degrees Fahrenheit as found according to the procedure on page 366.

TABLE 9-2 PERIMETER HEAT FLOW FRACTIONS

Construction	P per surface
Concrete floor slab, rigid insulation under and around edges	0.93
Masonry wall, rigid insulation inside, outside, or between wythes	0.95
Concrete block wall, cavities filled with loose insulation	0.68
Wood stud wall, batt insulation in cavities	0.77
Above, rigid insulation on outside	0.83
Metal stud wall, batt insulation in cavities	0.80
Above, rigid insulation on outside	0.87
Stud walls, cavities empty, rigid insulation on outside	0.93
Stud wall with much glass or door area	0.3
Above, rigid insulation on outside	0.5
Glass curtain walls	1.2
Concrete roof, rigid insulation on top	0.93
Wood or metal frame roof, batt insulation in cavities	0.91
Above, rigid insulation on top	0.94
Roof or ceiling, insulated, with much skylight, flue, or door area	0.42
Construction with no insulation	1.00

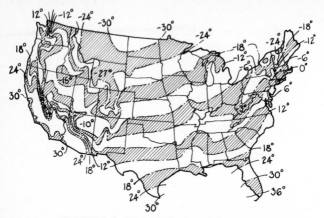

FIGURE 9-3 Winter design-temperature map.

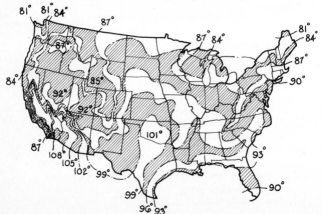

FIGURE 9-4 Summer design-temperature map.

1. For all calculations, find the local winter or summer design temperature from Fig. 9-3.
2. Adjust the design temperature according to the following conditions:
   a. In cold-weather calculations, if the envelope surface exists above grade:

$$T_o = T_i - W_c(T_i - T_d)$$

T_o = outdoor temperature
W_c = windchill exposure factor (See Table 9-3)
T_i = indoor temperature
T_d = design temperature

TABLE 9-3 WINDCHILL EXPOSURE FACTORS

Condition	Exposure factors (Fig. 9-4) *
Envelope surface fully shielded: Solid windscreen over whole surface, no diagonal winds	
Envelope surface mostly shielded: Few openings in surrounding windscreen, few diagonal winds	
Envelope surface partly shielded: ±50% gaps in surrounding windscreen	

TABLE 9-3 WINDCHILL EXPOSURE FACTORS *(Continued)*

Condition	Exposure factors (Fig. 9-4) *
Envelope surface mostly exposed: Little protection from direct or diagonal winds	
Envelope highly exposed: Facade or roof on hill or upper part of tall building	

*Exposure factors on side of buildings are for exterior walls; those on top of buildings are for roofs.

   b. In cold-weather calculations, if the envelope surface exists below grade:

     (1) Walls down to 16 ft below grade:

$$T_o = T_d + \left(\frac{A + T_d}{2}\right)\left(\frac{W_d}{16}\right)$$

     (2) Parts of floors to 16 ft down and in:

$$T_o = T_d + \left(\frac{A + T_d}{2}\right)\left(\frac{16 - W_d}{16}\right)$$

     (3) Wall and floor areas more than 16 ft down and in:

$$T_o = A$$

T_o = outdoor temperature
T_d = design temperature
 A = average annual temperature for the region (as listed in Fig. 9-5)
W_d = depth of wall, to 16 ft below grade, measured down and in (Fig. 9-6)

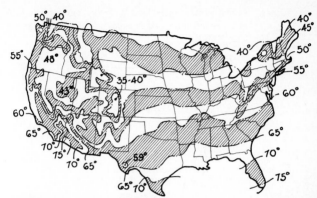

FIGURE 9-5 Average annual temperature map. *(From Robert Brown Butler, The Ecological House, Morgan & Morgan, 1981.)*

16 FEET DOWN & IN, MEAS-
URED ALONG OUTSIDE OF
WALL BELOW GRADE & UN-
DERSIDE OF BASEMENT
FLOOR, NOT INCLUDING
FOOTING

FIGURE 9-6 Section through wall and floor below grade.

c. In hot-weather calculations, if the envelope surface exists above grade and is exposed to direct sunlight at least 2 hr daily:

$$T_o = T_d + UIC\left(\frac{1200}{6R + W}\right) - L_o$$

T_o = outside temperature

T_d = design temperature

U = umbra fraction (portion of envelope surface that is unshaded during the day, maximum value during the day), as below:

> Surface unshaded all day, 1.0
> Surface shaded all day, 0.0
> Intermediate values obtained from observing site condition

I = incidence factor (portion of sunshine falling on envelope surface if it varies from vertical or due south, as shown in Fig. 9-7)

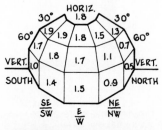

FIGURE 9-7 Incidence factors.

C = color coefficient, depending on color of envelope surface, as below:

Black or very dark, 1.0
Medium grey, 0.8
Glass, 0.7
White or very light, 0.5

R = thermal resistance of envelope construction, as explained above

W = weight of envelope construction in pounds per square foot, as listed in Table 3-1, p. 54

L_o = latitude-orientation factor (adjustment for latitude and orientation of envelope surface, as listed in Table 9-4)

TABLE 9-4 LATITUDE-ORIENTATION FACTORS

Latitude, degrees N.	Orientation*					
	S	SE/SW	E/W	NE/NW	N	Horiz.
24	−6	−3	0	2	1	1
32	−3	−1	0	1	1	1
40	1	0	0	0	0	1
48	4	3	1	0	0	0

*Interpolate for intermediate values.

d. In hot-weather calculations, if the envelope surface exists above grade and is not exposed to sunlight:

$$T_o = T_d$$

e. In hot-weather calculations, if the envelope surface exists more than one foot below grade:

$$T_o = \frac{T_i + AA}{2}$$

f. In all calculations, if an unheated void exists between the indoor space and outdoors:

Cold weather:

$$T_u = T_i - (T_i - T_o) \frac{R_i}{R}$$

Hot weather:

$$T_u = T_i + (T_o - T_i) \frac{R_i}{R}$$

T_u = temperature of the unheated void 1 ft outside the surface between unheated void and outdoors (this is the final "outdoor" temperature)

R_i = R value of insulation between unheated void and indoors

R = R value of total construction on both sides of unheated void

T_i = indoor temperature

T_o = outdoor temperature

t = timespan of heat flow in hours; for design temperature calculations, $t = 1$; for energy-auditing calculations, $t =$ number of hours in the auditing period (e.g., if the auditing is for the month of January, $t = 24 \times 31 = 744$ hr)

Example 1: Winter Heat Loss   A two-story residence near Albany, New York, is 40 ft long and 24 ft wide. The living room on the second floor is 13 x 17 ft in size and has a flat ceiling with an attic and a low-pitched gable roof above. The ceiling joists contain 10-inch nominal batt insulation, and the roof contains 1-in styrofoam between the sheathing and the shingles. If the house is nestled into a low hillside that rises to the north and west (the direction of prevailing winter winds), and these slopes are forested with tall evergreen trees, what is the design conduction heat loss through the living room ceiling? (See Fig. 9-8.)

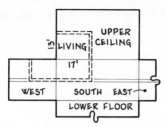

FIGURE 9-8 Foldout of part of residential building envelope.

$$C_c = \frac{A}{RP}(T_i - T_o)t$$

C_c = total conduction heat loss through the envelope construction, **?** BTU/hr

A = surface area of envelope (area of ceiling), $13 \times 17 = 221$ ft^2

R = R value of envelope construction (as listed in Table 9-1), $R = 3 + r_1 t_1 + r_2 t_2 + \ldots r_n t_n$

 r_1 = R value of batt insulation in ceiling (from Table 9-1), 3.5
 t_1 = thickness of batt insulation in ceiling, 9.5 in
 r_2 = R value of styrofoam insulation in roof (from Table 9-1), 4.2
 t_2 = thickness of styrofoam insulation in roof, 1 in

$$R = 3 + 3.5 \times 9.5 + 4.2 \times 1 = 40.5$$

P = perimeter heat flow fraction (as listed in Table 9-2, P for frame roof, batt insulation in cavities, rigid insulation on top), 0.94

T_i = indoor temperature, use 68°F

T_o = outdoor temperature:

1. Design temperature for region (from Fig. 9-3), $-3°$
2. Envelope surface exists above grade:

$$T_o = T_i - W_c(T_i - T_d)$$

T_o = outdoor temperature, ?°F
T_i = indoor temperature, 68°F
W_c = windchill exposure factor (from Table 9-3), 1.03
T_d = design temperature from above, -3°

$$T_o = 68 - 1.03[68 - (-3)] = -5°$$

3. Unheated void exists between indoors and outdoors:

$$T_u = T_i - (T_i - T_o)\frac{R_i}{R}$$

T_u = unheated void temperature, ?°F
T_i = indoor temperature, 68°F
T_o = outdoor temperature from above, -5°F
R_i = R of inner construction (ceiling); this equals R of batt insulation in ceiling plus half of 3:

$$3.5 \times 9.5 + 0.5 \times 3 = 34.8$$

R = R of total construction, 40.5

$$T_u = 68 - [68 - (-5)]\frac{34.8}{40.5} = 68 - 63 = 5°F$$

$t = 1$

$$C_c = \frac{221}{44 \times 0.94}(68 - 5) = 337 \text{ BTU/hr}$$

Example 2: Summer Heat Gain The west facade of an office building in Dallas, Texas, is 72 ft wide and 98 ft tall from ground level to the roof. Behind the facade on all upper floors are office spaces approximately 12 ft wide. The windows in each office extend from one side to the other and have their sills at 2 ft 6 in, heads at 7 ft, and white permanently positioned louvers mounted outside which keep the glass in minimum two-thirds shade. The weight of the window construction equals about 10 lb/ft^2. Under these conditions, what is the design cooling load for each office space through its windows? (See Fig. 9-9.)

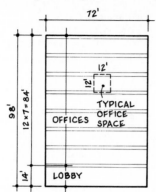

FIGURE 9-9   Foldout of part of office building envelope.

$$C_h = \frac{AF}{RP}(T_o - T_i)t$$

C_h = total conduction heat gain through the envelope construction, **?** BTU/hr

A = surface area of envelope (area of glass), $(7.0 - 2.5) \times 12 = 54$ ft^2

F = fan factor (no fan ventilation within the envelope materials), 1.0

R = R value of glass (from Table 9-1), 1.7

P = perimeter heat flow fraction (from Table 9-2, P for glass curtain walls), 1.2

T_o = outdoor temperature:

1. Design temperature for region from Fig. 9-3 = 99°
2. Envelope surface is exposed to direct sunlight at least 2 hr daily (in late afternoon); louvers keep two-thirds of surface in shade:

$$T_o = T_d + UIC \frac{1200}{6R + W} - L_o$$

T_o = outdoor temperature, ?°F
T_d = design temperature from above, 99°F
U = umbra fraction (one-third of glass exposed to sun), 0.33
I = incidence factor (from Fig. 9-6, I for vertical west facade), 1.5
C = color coefficient: one-third glass = 0.7, two-thirds white = 0.5

$$C = 1/3 \times 0.7 + 2/3 \times 0.5 = 0.57$$

R = R value of glass from above, 1.7
W = weight of envelope construction 10 lb/ft²
L_o = latitude-orientation factor (from Table 9-4, L_o for Dallas, 32° west), 0

$$T_o = 99 + 0.33 \times 1.5 \times 0.57 \times \frac{1200}{6 \times 1.7 + 10} - 0$$
$$= 99 + 17 = 116°F$$

T_i = indoor temperature, use 78°F
t = timespan at design load condition, 1 hr

$$C_h = \frac{54 \times 1}{1.7 \times 1.2} (116 - 78) \times 1$$
$$= 1006 \text{ BTU/hr}$$

INFILTRATION

Infiltration is the heat gained or lost through all open areas in the building envelope, such as seams, cracks, flues, and vent openings. This heat flow is represented by the following formulas.

Cold weather:

$$I_c = 0.018VAW_c(T_i - T_o)t$$

Hot weather:

$$I_h = 0.074VA(H_o - H_i)t$$

I_c = total infiltration heat loss for a room with a floor, wall, or ceiling surface in the building envelope during cold weather, in British thermal units (BTU) per hour

I_h = total infiltration heat gain for a room with a floor, wall, or ceiling surface in the building envelope during hot weather, in BTU per hour

V = volume of interior space behind the building envelope in cubic feet, as described in Table 9-5

TABLE 9-5 AIR CHANGES PER HOUR FOR ARCHITECTURAL VOLUMES*

Type of space	Single glass, no weatherstrip	Storm sash or weatherstrip
1. No windows or exterior doors	0.7	
2. One surface exposed to outdoors, average window or door area	1.2	0.8
3. Two surfaces exposed to outdoors, average window or door area	1.6	1.1
4. Three surfaces exposed to outdoors, average window or door area	1.9	1.3
5. Entrance halls	2.0	1.4
6. Large rooms with small openings	multiply above figures by 0.9	
7. Small rooms with large openings	multiply above figures by 1.2	

TABLE 9-5 *(Continued)*

Type of space	Single glass, no weatherstrip	Storm sash or weatherstrip
8. Ceilings or outer walls having plaster interior finish, stucco exterior finish, or poly. vapor barrier	multiply above figures by 0.4	
9. Spaces having exterior walls with all seams and utility penetrations sealed and / or top floor ceilings with insul. on top of interior walls and no recessed illumination	multiply above figures by 0.7	
10. Unopenable glass, awning or casement windows	multiply above figures by 0.8	
11. Rooms having HVAC ductwork outside the bldg. envelope	add 0.3 to above figures	
12. Fireplace:		
a. Loose damper	add 2500 / V to above	
b. Tight damper	add 1750 / V to above	
c. Glass doors	multiply no. 12 *a* or *b* by 0.7, then add to above	
13. Wood stove:		
Nonairtight	add 1500 / V to above	
Airtight	add 300 / V to above	

*When using this table for infiltration volume calculations, count only the top floor, volumes above grade exposed to the outdoors or unheated areas, and bottom floor spaces whose undersides are exposed to the outdoors or unheated voids. Do not count volumes under the top floor, volumes existing more than 14 ft in from roofs or exterior walls, and volumes whose horizontal projections exist more than 1 ft below grade. This is illustrated in Fig. 9-10.

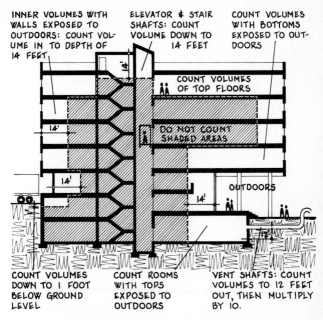

INNER VOLUMES WITH
WALLS EXPOSED TO
OUTDOORS: COUNT VOL-
UME IN TO DEPTH OF
14 FEET

ELEVATOR & STAIR
SHAFTS: COUNT
VOLUME DOWN TO
14 FEET

COUNT VOLUMES
WITH BOTTOMS
EXPOSED TO OUT-
DOORS

14'

COUNT VOLUMES
OF TOP FLOORS

14'

DO NOT COUNT
SHADED AREAS

14'

OUTDOORS

14'

COUNT VOLUMES
DOWN TO 1 FOOT
BELOW GROUND
LEVEL

COUNT ROOMS
WITH TOPS
EXPOSED TO
OUTDOORS

VENT SHAFTS: COUNT
VOLUMES TO 12 FEET
OUT, THEN MULTIPLY
BY 10.

FIGURE 9-10 Building air changes per hour.

A = number of air changes per hour for the type of space behind the building envelope, as described in Table 9-5

W_c = Windchill exposure factor, from Table 9-3, p. 365

T_i = indoor temperature, usually 65 to 72° F in winter

T_o = outdoor temperature (calculate from pp. 366–370)

H_i = enthalpy of indoor air during hot weather, in BTUs per pound; when temperature and relative humidity are known, enthalpy may be found from Fig. 9-12, the summer design temperature for a region may be found from Fig. 9-4, and the summer design humidity from Fig. 9-11; for air at 78°F and 50 percent r.h. (an often-used indoor design condition), $H = 30.1$ BTU/lb ≈ 30 BTU/lb

H_o = enthalpy of outdoor air during hot weather (as found from outdoor design temperature, Fig. 9-3, relative humidity, Fig. 9-11, and enthalpy, Fig. 9-12)

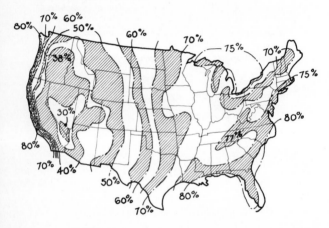

FIGURE 9-11 Summer relative humidity map.

t = timespan of infiltration heat flow in hours (for design temperature calculations, $t = 1$); for energy auditing calculations, $t =$ number of hours in the auditing period (e.g., if the audit is for the month of February, $t = 24 \times 28.5 = 684$ hr)

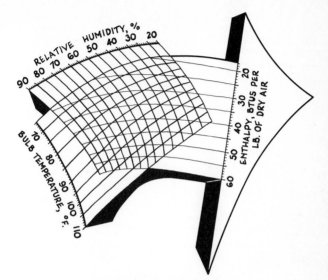

FIGURE 9-12 Enthalpy graph.

Example 1: Winter Heat Loss A house near Albany, New York, has a 13 x 17 ft living room on the southeast corner of the second floor. The space is 8 ft high. The room has a small openable window in the west wall, a large area of fixed glass in the south wall, a polyethylene vapor barrier between the framing and interior finish on the outer walls and ceiling, sealed framing seams and utility penetrations, insulation over all interior walls, no recessed illumination, and an airtight wood stove with damper. What is the design infiltration heat loss of this space?

$$I_c = 0.018VAW_c(T_i - T_o)t$$

I_c = total infiltration heat loss for the space, ? BTU/hr
V = volume of the space, $13 \times 17 \times 8 = 1768$ ft^3
A = number of air changes per hour, from Table 9-5:

> type 4: three surfaces exposed to outdoors, openings weatherstripped, 1.3
>
> type 8: ceilings and outer walls have vapor barrier, $1.3 \times 0.4 = 0.56$
>
> type 9: space has all seams sealed, etc, $0.56 \times 0.7 = 0.39$
>
> type 13: wood stove, airtight, add $300/V$ to above:

$$\text{Total } A = 0.39 + \frac{300}{V} = 0.39 + \frac{300}{1768} = 0.56$$

W_c = Windchill exposure factor, assume average conditions; $W_c = 1.1$
T_i = indoor temperature, use 68°F
T_o = outdoor temperature (use same outdoor temperature as calculated in Example 1 under Conduction, -5°F)
t = 1 hr

$$I_c = 0.018 \times 1768 \times 0.56 \times 1.1 \times [68 - (-5)] \times 1$$
$$= 1431 \text{ BTU/hr}$$

Example 2: Summer Heat Gain In a Dallas, Texas, office building, a typical office space measures 11 ft 4 in from floor to floor, 12 ft wide, and 16 ft deep. The windows are fixed pane and their frames are solidly caulked. What is the summer heat gain due to infiltration for one office space having its side walls common to adjoining offices and its outer wall in the building envelope?

$$I_h = 0.074VA(H_o - H_i)t$$

I_h = total infiltration heat gain for the space, ? BTU/hr
V = volume of the space:

> 11 ft 4 in high from floor to floor = 11.33 ft
> 12 ft wide

16 ft deep (count only the outer 14 ft)
11.33 × 12 × 14 = 1903 ft³

A = number of air changes per hour, from Table 9-5:

> type 2: one surface exposed to outdoors, openings weather-
> stripped, 0.8
> type 10: unopenable glass, 0.8

$$\text{Total } A = 0.8 \times 0.8 = 0.64$$

H_o = outdoor enthalpy (depends on design temperature and humidity
for Dallas):

> From Fig. 9-4, Dallas design temperature = 99°
> From Fig. 9-11, Dallas design humidity = 65%
> From Fig. 9-12, enthalpy at 99° and 65% = 53 BTU/lb

H_i = indoor enthalpy (use indoor temperature of 78°F and humidity
of 50%: H at 78°F and 50%), 30 BTU/lb

t = 1 hr

$$I_h = 0.074 \times 1903 \times 0.64(53 - 30) \times 1$$
$$= 2070 \text{ BTU/hr}$$

AUXILIARY HEAT GAIN

Auxiliary heat gain is the amount of heat added to interior spaces by
people, lighting, appliances, and machines. During hot weather, maximum auxiliary heat gains and cooling loads often occur at the same
time (midafternoon); thus, the former are usually added to the latter
when sizing cooling equipment. However, during cold weather the coldest time of day is usually just before sunrise, a time when auxiliary heat
gains may be at a minimum. Thus, when calculating design heat loads,
consider only the auxiliary heat sources that exist from 6 to 8 a.m., usually the time of the maximum heating load.

Auxiliary heat gain is described by the formula below.

$$A_h = [NM + 0.034W(100 - 0.6E) + H]t$$

A_h = total heat gain from all auxiliary heat sources present in the space at the time of design condition (usually 3 p.m. for cooling and 7 a.m. for heating), BTU/hr

N = number of occupants in the space at the time of design heating or cooling condition

M = Average metabolism of each occupant according to activity, in BTU/hr, as listed in Table 9-6

TABLE 9-6 METABOLISM OF HUMAN ACTIVITIES

Activity	Metabolism, BTU/hr		
	Male	Female	Child
Sleeping	250	210	190
Seated:			
Quiet, as in theatre	360	300	270
Writing, light work	480	410	360
Eating (includes food heat)	520	450	400
Typing or clerical	640	540	480
Standing:			
Slow walking	800	680	600
Light machine work	1000	850	750
Heavy work, lifting	1600	1300	1100
Walking, 3 mi/hr	1000	850	750
Jogging, gym athletics	2000	1700	1500

W = total wattage of illumination used in space at the time of design heating or cooling condition, in watts

E = efficacy of lamps in the space (as listed in Table 7-1)

H = heat gains from motors, machinery, and appliances located in the space; for electrically operated machines and appliances, use the following formula, which is a general estimate:

$$H = 1.1VA$$

H = heat gain of the motor or appliance in BTU/hr
V = nameplate voltage
A = nameplate amperage

For nonelectrical machinery, specific calculations relating to the heat generated by each machine should be performed, or else one may use the general area auxiliary heat gain rates as listed in Table 9-7

TABLE 9-7 AUXILIARY HEAT GAIN RATES FOR MACHINERY

Type of space	Heat gain, BTU/hr per ft²
Residences	1
Apartments	1.5
Hospitals	2
Offices:	
General	3–4
Purchasing and accounting departments	6–7
Computer display areas	12–15
Laboratories	15–70
Restaurants:	
Eating areas	3
Kitchen areas	150–200
Computer areas	50–150
Manufacturing:	
General assembly and stamping	20
Planting, forming, curing	150

t = time span of auxiliary heat gain, in hours (for design temperature calculations, $t = 1$); for energy auditing calculations, t = number of hours in the auditing period (e.g., if the audit is for the month of November, $t = 24 \times 30 = 720$ hr)

Example 1: Winter Condition   What is the auxiliary heat gain for design winter conditions in a residence for a family of four near Albany, New York?

$$A = [NM + 0.034W(100 - 0.6E) + H]t$$

A = total auxiliary heat gain (design winter condition, taken at 7 a.m. for the residence), **?** BTU/hr

N = number of occupants, 4

M = average metabolism of the occupants (from Table 9-6, at 7 a.m., occupants are probably sleeping), from Table 9-6:

      Adult male, sleeping, 250   250

      Adult female, sleeping, 210   210

      Two children, sleeping, 190 $\times$ 2 = $\underline{380}$

                               $\overline{840}$ ÷ 4 = 210 BTU/hr

W = wattage of illumination used during design condition; at 7 a.m. total wattage used is probably zero; thus, $0.034W(100 - 0.6E) = 0$

H = heat gains from motors, machinery, and appliances located in the space; at 7 a.m. probably zero

t = 1 hr

$$A = 4 \times 210 - 0 = 840 \text{ BTU/hr}$$

Example 2: Summer Condition   What is the auxiliary heat gain for design summer conditions in a 13 x 17 office in a building in Dallas, Texas? The space is used by two persons and has the usual office machinery in it. Illumination is four ceiling-mounted fluorescent fixtures, each containing four 40-W lamps.

$$A = [NM + 0.034W(100 - 0.6E) + H]t$$

A = total auxiliary heat gain (design summer condition, taken at 3 p.m., for the office space), **?** BTU/hr

N = number of occupants, 2

M = metabolism of occupants (from Table 9-6; two adult males doing writing, light work), 480 BTU/hr

W = total wattage of illumination in office (four fixtures with four 40-W lamps each), $4 \times 4 \times 40 = 640$ W

E = efficacy of the illumination (as listed in Table 7-1, E for 40-W fluorescent lamp), 75

H = heat gains from motors, machinery, and appliances in the space (from Table 9-7, H for general office space = 3.5 average BTU/ hr per square foot), total $H = 3.5 \times 221 = 774$ BTU/hr

t = 1 hr

$A = [2 \times 480 + 0.034 \times 640(100 - 0.6 \times 75) + 774] \times 1$
$= 2930$ BTU/hr

PICKUP HEATING LOAD

Pickup heating load is a winter condition in which heat is absorbed by interior solid materials that have become cold while the heating system has been off for an extended time. In intermittently heated architecture, this situation can significantly increase the design heating load.

Pickup heating loads vary somewhat according to the type of furnishings and construction inside the building envelope, the amount of surfaces exposed to moving air, and other factors. Thus, the calculations in this section are general approximations.

Pickup heating load is described by the following formula.

$$P = H(C + I)$$

P = pickup heating load of a space in British thermal units (BTU) per hour

H = pickup heating load fraction (portion of heat load absorbed by interior solid materials depending on intermittence of heat flow, as listed in Table 9-8)

TABLE 9-8 PICKUP HEATING LOAD FRACTIONS

Type of heat flow	P
Heat off for 6 hr, or nightly setback of 10°F	0.1
Heat off for 12 hr, or nightly setback of 20°F	0.2
Heat off for 18 hr	0.3
Heat off for 30 or more hr	0.5

C = design conduction heating load in BTU/hr
I = design infiltration heating load in BTU/hr

Example 1: Pickup Heating Load A small rural church in northeast Montana is used for services every Sunday morning. In winter the heat is turned on during the service. If the design conduction and infiltration heating loads are 118,000 BTU/hr and 96,000 BTU/hr, what is the building's pickup heating load?

$$P = H(C + I)$$

P = pickup heating load, ? BTU/hr
H = pickup heat load fraction (from Table 9-8, H for heat off 30 or more hours), 0.5
C = conduction heat load, 118,000 BTU/hr
I = infiltration heat load, 96,000 BTU/hr

$$P = 0.5(118,000 + 96,000)$$
$$= 107,000 \text{ BTU/hr}$$

SOLAR HEAT GAIN

Solar heat gain is the amount of heat added to interior spaces during winter by solar energy passing through glass or similar material existing in the southernmost surfaces of the building envelope. With certain kinds of architecture, in cold weather this radiance can be a dominant source of heat supply. Solar heat gain rarely affects the sizing of a building's mechanical or "backup" heating system, for winter design temperatures usually occur just before dawn when no incoming radiance is available to reduce design heating loads. However, solar energy use can greatly reduce the operation times—and thus fuel costs—of mechanical heating units.

Solar heat gain is represented by the formula below. This equation does not consider conduction heat flowing out through the glass as sunshine is flowing in.

$$S = AUTECI\left(\frac{P + PG}{100}\right)t$$

S = total heat gained from solar energy entering interior spaces through the building envelope in British thermal units per day

A = net area of glazing surface, not including frames or other architectural obstructions, in square feet

U = umbra fraction (portion of glazing surface that is unshaded, daily average):

Collector surface in shade all day, 0

Collector surface in sun all day, 1.0

Intermediate values obtained from observing conditions on the site

T = transmittance fraction of the glazing material (the portion of solar energy passing completely through the glazing, as listed in Table 9-9)

TABLE 9-9 TRANSMITTANCE FRACTIONS FOR GLASS,*

Type of glazing	T
Single pane:	
Clear, low-iron glass	0.92
Clear, standard glass	0.88

TABLE 9-9   (Continued)

Type of glazing	T
Heat-absorbing glass	0.62
Reflective glass	0.40
Double pane:	
Clear-clear (low-iron)	0.85
Clear-clear (standard)	0.78
Heat-absorbing-clear (standard)	0.55
Reflective-clear (standard)	0.35
With standard insect screen over	multiply above by 0.7

*Rigid plastics are usually unsatisfactory for waterproof solar glazing because of their high rates of thermal expansion.

E = amount of solar energy incident upon a vertical surface facing due south, in BTUs per square foot of glazing per clear day (depends on latitude and month of year, as listed in Table 9-10)

TABLE 9-10   INSOLATION DATA, BTU per clear day*

Latitude, °	Nov. 21	Dec. 21	Jan. 21	Feb. 21	Mar. 21
24	1614	1686	1647	1377	953
32	1624	1673	1659	1533	1190
40	1572	1535	1610	1614	1384
48	1345	1216	1378	1604	1522

*Incident solar energy on a vertical surface facing due south, representative values for the 21st of each month.

C = clarity fraction (clearness of local atmosphere, depending on location, elevation above sea level, smog, and other environmental factors, as listed in Table 9-11)

TABLE 9-11 CLARITY FRACTIONS FOR LOCAL ATMOSPHERES

Region or condition	Clarity fraction, winter months
Florida, Gulf coast, middle south, west coast	0.95
Deep south	0.90
East Canadian border, upper Great Lakes, northern plains, Rocky Mountains	1.05
All other areas in 48 states	1.00
6000 ft above sea level	add 0.05 to above
Continuous light smog or local haze	subtract 0.1 from above
Continuous heavy smog	subtract 0.2 from above

I = incidence factor (portion of sunlight absorbed by collector surface if its orientation varies from vertical and due south, as shown in Fig. 9-13)

P = percentage of sunshine (average percentage of possible sunshine for the region's coldest months of the year, from local climatic data)

G = ground reflectance factor (amount of solar energy reflected onto the collector surface from the ground in front); the formula below should be used only if the collector surface is vertical or nearly so, its bottom is at or near the ground, and the ground is fairly level 50 feet in front, otherwise, $G = 0$

$$G = A[B + C(0.7 - B)]$$

A = aspect of smoothness of ground, as listed in Table 9-12

B = bare ground reflectance, as listed in Table 9-12

C = snow cover reflectance, as listed in Table 9-12

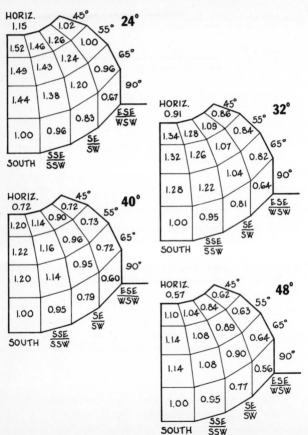

FIGURE 9-13 Incidence factors: tilt of collector vs. orientation from due south.

TABLE 9-12 GROUND REFLECTANCE FACTORS

Nature of ground just outside collector surface	Reflectance factors
ASPECT OF SMOOTHNESS, A	
Ground surface:	
Fairly smooth	1.0
Rolling, varied, or slightly rocky	0.5
BARE GROUND REFLECTANCE, B	
Dirt surface:	
Light color	0.25
Dark color	0.1
Vegetation:	
Yellow or light green	0.25
Medium to dark green	0.1
Concrete, gravel, yellowish or tan sand	0.25
Sand, white or near-white	0.4
Red brick or weathered (dark) wood terracing	0.13
Asphalt, slate	0.05
White smooth surface	0.7
SNOW COVER REFLECTANCE, C	
Less than 3 days per month	0.0
Average per month:	
5 days	0.1
10 days	0.2
15 days	0.3
20 days	0.4
25 days	0.5
30 days	0.6

t = timespan of heat gain, in days (for weekly, monthly, or seasonal calculations, t equals the number of days in the time period)

Example   A two-story residence 40 ft long and 24 ft wide near Albany, New York, has two-story tall thermopane glass on the south facade for collecting solar energy. The facade is vertical, faces southsoutheast, and is shaded in summer by three large maple trees whose trunks are 20 to 30 ft away. From this facade extends a broad, smooth lawn which in January is covered with snow an average 25 days per year. On a typical January day, how much solar energy passes through this glass? A drawing of the south elevation is shown in Fig. 9-14.

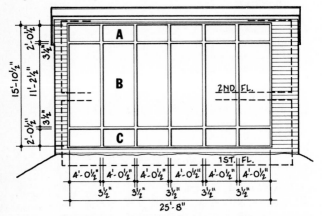

FIGURE 9-14   South elevation of solar house.

From specifications, net glass area (do not count frame or trim)

Pane areas A & C: 5.89 ft² each (43 in × 19¾ in pane)
Pane areas B: 43.4 ft² each (47 in × 121 in pane size)

$$S = AUTECI\left(\frac{P + PG}{100}\right)t$$

S = total heat gain from solar energy passing through the glass in BTUs per day (use January as design condition)

A = area of collector surface (net area of glass from specs and elevation drawing), $5.89 \times 2 \times 6 + 43.4 \times 6 = 331$ ft^2

U = umbra fraction (the portion of collector surface that is not shaded during the day; from site inspection, the bare maple tree branches shade about 20% of the collector surface during a clear winter day), use 0.8

T = transmittance fraction (from Table 9-9, T for double glazing, clear-clear), 0.78

E = amount of solar energy, in BTUs per day per square foot, incident upon a vertical surface on a clear day, depending on latitude; of Albany = $42°40'$; from Table 9-10:

Latitude	E in January
40°	1610
42°40′	? (interpolate)
48°	1374

Interpolation:

$$1610 - \frac{2.67}{8}(1610 - 1374) = 1530 \text{ BTU/day per ft}^2$$

C = clarity fraction for region (from Table 9-11; Albany is not near East Canadian border, thus C for all other areas in 48 states), 1.00

I = incidence factor (from Fig. 9-13, I for vertical surface facing SSE at 40° lat.), 0.95

P = average percentage of possible sunshine (from local climatic data, P for Albany for January), 46%

G = ground reflectance factor; collector surface is vertical, bottom is near ground, and landscape in front is nearly level, so use the following formula: $G = A[B + C(0.7 - B)]$

A = smooth ground surface (from Table 9-12), 1.0
B = lawn that is light green in winter (from Table 9-12), 0.25
C = 25 days snow cover per month (from Table 9-12), 0.5

$$G = 1.0[0.25 + 0.5(0.7 - 0.25)] = 0.45$$

t = 1 day of average January weather

$$S = 331 \times 0.8 \times 0.78 \times 1530 \times 0.95 \left(\frac{46 + 46 + 0.45}{100} \right)$$
$$= 200,000 \text{ BTU collected during an average January day}$$

Energy Savings Due to Solar Architecture

An often crucial aspect of solar heat gain analysis is determining the heating bills of a building equipped with a solar energy system. In this way, the system's cost-effectiveness can be established. This is done in the following manner.

STEP 1.   Find the solar load ratio (SLR) of the architecture's solar heating system.

$$\text{SLR} = \frac{S}{C + I - A + P}$$

SLR = solar load ratio (the ratio of solar heat gain to design heat loss through the building envelope)
S = solar heat gain, from page 387 (see Note below)
C = conduction heat loss, from page 361 (see Note below)
I = infiltration heat loss, from page 374 (see Note below)
A = auxiliary heat gain, from page 381 (see Note below)
P = pickup heating load, if any, from page 385 (see Note)

NOTE: These are not design (i.e., peak) heating loads but are average heating loads for an extended period of time, usually 1 month. These calculations are performed as described earlier, except that T_o equals the average outdoor temperature for the month as found in local climatic data, and t equals the number of hours or days in the month. All other unknowns in the solar equations are monthly averages, with solar trajectory information for the 21st of the month being considered as the average. By calculating heating loads for every winter month, seasonal heating loads can be determined.

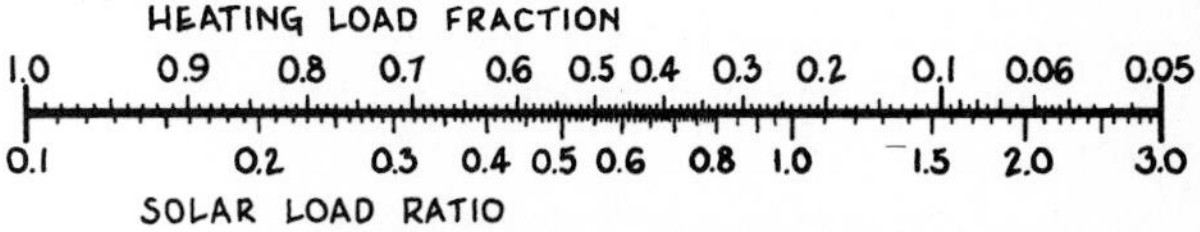

FIGURE 9-15   SLR vs. HLF bar graph.

STEP 2.   Knowing the SLR, from Fig. 9-15 find the heating load fraction (HLF) resulting from solar energy use.

STEP 3.   Determine the cost of heating the solar architecture from the formula below.

$$\text{Cost} = \frac{\text{HLF}(C + I - A + P)K}{EF}$$

Cost = cost of heating the solar architecture for the period of the energy audit

HLF = heating load fraction from step 2

  C = conduction heat loss from step 1 in BTU

  I = infiltration heat loss from step 1 in BTU

  A = auxiliary heat gain from step 1 in BTU

  P = pickup heating load from step 1 in BTU

  K = unit cost of energy from local data

  E = energy conversion factor, depending on type of energy used by conventional or backup heating system, from Table 10-1

  F = efficiency of backup heating system, from Table 9-13

TABLE 9-13 EFFICIENCY OF HEATING SYSTEMS, F

Type of heating system	F
Electric baseboard heaters, portable electric heaters	0.9
Furnace with air ducts or water baseboards	0.65
Kerosene space heater	0.9
Fireplace, open hearth	0.2
Fireplace with glass doors or nonairtight wood stove	0.4
Wood stove or coal stove, airtight	0.6

Example A superinsulated solar residence near Albany, New York, has the following average heat load statistics for the month of January.

Average daily solar heat gain, 200,000 BTU/day
Average conduction heat load, 2,234 BTU/month
Average infiltration load, 2,412 BTU/month
Average auxiliary heat load, 156,000 BTU/month
Average pickup heat load, 0

If this house is warmed by electric resistance baseboard heaters and the cost of electricity is 7.2 cents per kilowatthour, what is the house's projected January heating bill?

STEP 1. Find the solar load ratio.

$$\text{SLR} = \frac{S}{C + I - A + P}$$

SLR = solar load ratio for the month of January, ?
 S = average solar heat gain for January, 200,000 BTU/day × 31 days = 6,200,000 BTU/month
 C = average conduction heat load for January, 2,234,000 BTU
 I = average infiltration heat load for January, 2,412,000 BTU
 A = average auxiliary heat load for January, 156,000 BTU
 P = average pickup heating load for January, 0

$$SLR = \frac{6,200,000}{2,234,000 + 2,412,000 - 156,000} = 1.38$$

STEP 2. Find the heating load fraction from Fig. 9-15. From Fig. 9-15, when SLR = 1.38, HLF = 0.13.

STEP 3. Determine the house's cost of heating for the month of January.

$$Cost = \frac{HLF(C + I - A + P)K}{EF}$$

Cost = cost of heating the house for January, ? dollars
HLF = heating load fraction from step 2, 0.13
  C = conduction heat load from step 1, 2,234,000 BTU
  I = infiltration heat load from step 1, 2,412,000 BTU
  A = auxiliary heat load from step 1, 156,000 BTU
  P = pickup heating load from step 1, 0
  K = unit cost of energy, 7.2 cents/kWh = 0.072 dollars/kWh
  E = energy conversion factor (electricity, from Table 10-1), 3413
  F = efficiency of electric baseboard heater (from Table 9-13), 0.9

$$Cost = \frac{0.13(2,234,000 + 2,412,000 - 156,000) \times 0.072}{3413 \times 0.9}$$

= \$13.68 heating cost for month of January

NOTE: This amount is extremely low for two reasons. First, the house's building envelope is superinsulated and thus its basic heating losses are low. Second, the little heat loss remaining is mostly replenished by the efficient solar energy system.

SUMMARY FOR DESIGN CONDITIONS

Total heating and cooling loads for design (i.e., peak) conditions are summarized by the following formulas.

Total heating load:

$$T_h = C + I - A + P$$

Total cooling load:

$$T_c = C + I + A$$

T_h = total heating load at design temperature in BTU per hour
T_c = total cooling load at design temperature in BTU per hour
C = design conduction heating or cooling load in BTU per hour
I = design infiltration heating or cooling load in BTU per hour
A = design auxiliary heat gain or loss in BTU per hour
P = design pickup heating load in BTU per hour

Example 1 What is the total design heating load of a small church in northeast Montana if the design conduction heating load is 118,000 BTU/hr, infiltration heating load is 96,000 BTU/hr, and the pickup heating load is 107,000 BTU/hr?

$$T_h = C + I - A + P$$

T_h = total heating load, **?** BTU/hr
C = conduction heating load, 118,000 BTU/hr
I = infiltration heating load, 96,000 BTU/hr
A = auxiliary heat gain (none at 7 a.m.), 0
P = pickup heating load, 107,000 BTU/hr

$$\begin{aligned}
T_h &= 118,000 + 96,000 - 0 + 107,000 \\
&= 321,000 \text{ BTU/hr}
\end{aligned}$$

Example 2 What is the total design heating load of a solar house near Albany, New York, if the design conduction heating load is 8840 BTU/hr, infiltration heating load is 9560 BTU/hr, auxiliary heat gain is 840 BTU/hr, and the design solar heat gain is 185,000 BTU/day?

$$T_h = C + I - A + P$$

T_h = total heating load, **?** BTU/hr
C = conduction heating load, 8840 BTU/hr
I = infiltration heating load, 9560 BTU/hr
A = auxiliary heat gain, 840 BTU/hr
P = pickup heating load, 0

$$T_h = 8840 + 9560 - 840 + 0$$
$$= 17{,}600 \text{ BTU/hr}$$

NOTE: Although a solar energy system can greatly reduce heating bills (as was demonstrated under "Energy Savings due to Solar Architecture"), it rarely reduces the design heat load of solar architecture or the size of its backup heating system.

Example 3 What is the total design cooling load of a 13 x 17 ft office space in a building in Dallas if its design conduction cooling load is 1006 BTU/hr, infiltration cooling load is 2070 BTU/hr, and auxiliary heat gain is 2930 BTU/hr?

$$T_c = C + I + A$$

T_c = total cooling load, **?** BTU/hr
C = conduction cooling load, 1006 BTU/hr
I = infiltration cooling load, 2070 BTU/hr
A = auxiliary heat gain, 2930 BTU/hr

$$T_c = 1006 + 2070 + 2930$$
$$= 6006 \text{ BTU/hr}$$

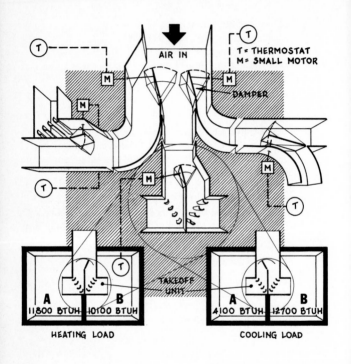

Ducting takeoffs.

CLIMATE CONTROL

SELECTING THE PROPER SYSTEM

During periods of heat imbalance between indoors and outdoors, climate control equipment in the architecture creates the opposite effect of the weather and keeps indoor temperatures comfortable.

The most important criteria for choosing a satisfactory climate control system are the design heating and cooling loads of the architecture and its owner's goals. The design heating and cooling loads are calculated as described in the previous chapter. The owner's goals may be:

1. Low initial cost
2. Low operating costs
3. Adequate level of performance that matches the competition
4. Ultimate in performance no matter what the cost

Table 10-1 presents energy conversion factors for several different types of fuel that can be used in a climate control system.

The positive and negative aspects of several climate control systems popularly used today are described below.

Electric Heating

+ Compact, easily installable units. Noiseless, odorless. Doesn't require fuel storage, combustion air, flues, ducts, or furnace space.
− Does not humidify, filter, or ventilate air. Operating costs are high in areas that have high electricity rates.

Electric Heating and Cooling

+ Through-the-wall packages that are easy to install and require only an electrical outlet to run. They heat, cool, humidify, filter, and freshen air year-round. Useful in architecture that has many rooms requiring individual climate control.

TABLE 10-1 ENERGY CONVERSION FACTORS, E

Fuel	Coal, lb	Oil, gal	Gas, therm	Electricity, kWh	Oak, lb	Solar, square
Coal, lb	14,600 BTU	0.160	0.195	4.28	3.73	0.487
Oil, gal	6.23	91,000 BTU	1.21	26.7	23.3	3.03
Gas, therm (100 ft³)	5.14	0.824	75,000 BTU	22.0	19.2	2.50
Electricity, kWh	0.234	0.038	0.046	3413 BTU	0.873	0.114
Wood, 1 lb dry	0.268	0.043	0.052	1.15	3910 BTU	0.130
Solar, square (100 ft²)	2.05	0.330	0.40	8.79	7.67	30,000 BTU

   —   Not good in large spaces. Operating costs are high in areas that have high electricity rates.

Hot Water Heating

   +   Works well in residences and small buildings that have high heating and low cooling loads. Little space or extra construction is required to house the plumbing.

   —   Central heating system takes up considerable space and performs no humidifying, filtering, or ventilating. Plumbing must be protected from freezing in winter and condensation in summer.

Steam Heating

   +   The steam can be used for other purposes, such as cooking, sterilizing, operating industrial processes, and driving generators.

   —   Design involves advanced engineering analysis and extensive safety apparatus. Rarely used today; will not be considered further in this chapter.

Air Heating

   +   Practical in homes and other small structures in cold climates. Can ventilate, humidify, and filter incoming air.

   —   Bulky furnace, ducting, and chimney cost a lot and take up much architectural space.

Air Cooling

   +   Compact packages are easily installed in walls or roofs of building envelope. Cooling capacities vary greatly, from ⅛ to 50 or more tons. Can humidify, filter, and freshen air. Some units can have small furnaces installed in them for heating in winter. Best for large, one-floor buildings in warm climates.

   —   Not practical in regions that have mild summers.

Air Heating and Cooling

+   Versatile, centralized system. Most effective in regions that have widely variable climates and in large buildings with large inner volumes, multiple zones requiring constant temperature and humidity control, or large steady loads having modest ventilation requirements.
−   expensive, often custom-assembled on the site. Requires considerable balancing of components before it operates smoothly.

Air-Water Heating and Cooling

+   Versatile, centralized system. Especially suitable in large buildings that have reversing heat flow requirements (in which heating may be required in one space and cooling in an adjoining space) and in skyscrapers and other structures that have minimum space requirements for mechanical equipment.
−   Complicated, costly. Not practical unless the smaller volume of its feeder systems can appreciably lower architectural costs.

Heat Pump

+   One machine heats and cools economically where winter temperatures stay above 35°F for prolonged periods. Good where heating and cooling loads are about the same and electricity rates are low.
−   Not practical in large, multizoned buildings. Part of machinery must go indoors and part outdoors. Requires backup heating at temperatures below 35°F and cannot produce heat below 15°F.

ELECTRIC HEATING

Electric heating design involves determining the wattage of the heating unit for each space.

The *general recommendations* for an electric heating design are:

1. Electrical heating units operate on 120 or 240 V.
2. The lower voltage is more economical only in small spaces.
3. There are several types of electric heating units:
   a. Baseboards: simple, unobtrusive
   b. Floor inserts: good for locating under floor-length glass
   c. Spot: small units for small spaces that have limited wall or baseboard area
   d. Surface mounted: good against brick, stone, or other hard surfaces
   e. Semirecessed: desirable for soft-hard construction such as furring on masonry walls
   f. Recessed: good for fitting into stud walls
   g. Fan-driven: not silent, but circulate heat better; good for ceiling-mounted fixtures in garages and other larger unfinished or semi-finished spaces
   h. Radiant cables: nice for melting snow and ice off eaves and sidewalks
4. Recessed units may be more attractive, but they are usually energy inefficient.
5. Some electric heaters have the thermostat mounted in the unit.

Sizing the Electric Heating Unit

Example 1 The conference room in a suburban real estate office has a 12-ft long electric baseboard heater in it, but in very cold weather the room remains chilly. If the design heat load of the room is 10,200 BTU/hr, how much additional baseboard heating should this room have?

$$H = 3LR$$

H = design heating load of the space, 10,200 BTU/hr
L = total length of electric baseboard heating, ? ft
R = rating of electric baseboard heater per foot of length (listed in manufacturer's catalogs and on the heater package), usually about 175 W/lin ft

$$10,200 = 3 \times L \times 175$$
$$L = 19.4 \text{ ft total length}$$

Use 20 ft. Additional baseboard length:

$$20 - 12 = 8 \text{ ft}$$

Example 2 The men's room in a bus terminal has a design heat loss of 3870 BTU/hr. The owner wants to install an electric spot heater in the wall. What capacity should it have?

$$H = 3W$$

H = design heating load of the space, 3870 BTU/hr
W = capacity of the heating unit in watts (the wattage of any electric heating unit is listed on its package and on the unit itself), ? W

$$3870 = 3W$$
$$W = 1290 \ W \text{ minimum}$$

From product catalogs or examination of units, choose a heating unit whose capacity is 1290 W or greater.

Example 3 The entrance apron of a church near Philadelphia is to contain radiant heating to prevent ice from forming on it in winter. If the design temperature in Philadelphia is 10°F and it rarely snows more than 1 in/hr there, how much radiant heating is required in the entrance apron?

This problem is solved by using Table 10-2. To the right of the snow-fall rate of 0–1 in/hr and under the design temperature of 10°F, find the proper radiant slab output. The answer is 43 W/ft².

TABLE 10-2   RADIANT SLAB OUTPUT, W/ft^2

Design rate of snowfall, in/hr	Slab output at design temperature and 5 mi/h wind			
	30°	20°	10°	0°
0–1	25	35	43	51
1–2	46	55	64	74
2+	68	79	90	98

NOTE: Radiant electrical cable for heated concrete slabs comes in capacities of 10, 20, 40, and 60 W/ft^2. Even in severely cold climates, the 60-W cable is satisfactory most of the time. When it is not, a more elaborate system of plumbing flowing with antifreeze must be devised.

Radiant slabs can be designed with probe thermostats and moisture sensors that will turn the system on automatically when the slab is below 32°F and its surface is moist.

Radiant slabs must be able to drain easily.

ELECTRIC HEATING AND COOLING

Electric heating and cooling design includes:

1. Determining the capacity and number of units required
2. Selecting the unit location

The *general description and recommendations* are:

1. These systems are small window-mounted or through-the-wall packages that have no ducting.
2. Each unit heats, cools, filters, and ventilates the air in its zone.
3. Capacities range from about 6500 to 30,000 BTU/hr for cooling and to 35,000 BTU/hr for heating.

4. Special features in some units are heaters energized by hot water or steam, high ventilating apparatus for densely occupied spaces, and grilles with adjustable vanes.

5. All units have self-contained adjustable thermostats.

Capacity and Number of Units

The drafting room of a tool and die company has a design heating load of 57,000 BTU/hr, a cooling load of 44,500 BTU/hr, and low ventilation requirements. The room's two exterior walls contain four large windows whose sills are 40 in high, and the only energy available is electricity. What wattage should the heating/cooling units be?

PROCEDURE Find the minimum heating and cooling capacity for the units from the formulas below.

Heating by water or steam:

$$H = 0.87NC_h$$

Heating by electricity:

$$H = 4NW_h$$

Cooling by electricity:

$$C = 3NW_c$$

H = design heating load of the conditioned space, 57,000 BTU/hr
C = design cooling load of the conditioned space, 44,500 BTU/hr
N = number of units desirable or practical in the conditioned space; the best arrangement would be to locate one incremental unit under each window; thus, there are four units
C_h = heating capacity of each unit in BTU per hour, if it is heated by hot water or steam
W_h = heating capacity of each unit in watts, if it is heated by electricity, ? W
W_c = cooling capacity of each unit in watts, if it is cooled by electricity, ? W

In this case, use the second and third formulas. The one yielding the highest wattage determines the minimum wattage of each heating/cooling unit.

$$H = 4NW_h\,C \qquad\qquad C = 3NW_c$$
$$57,000 = 4 \times 4W_h \qquad 44,500 = 3 \times 4W_c$$
$$W_h = 3560\ \text{W} \ W_c \qquad W_C = 3710\ \text{W}$$

Highest wattage = 3710 = minimum wattage of each unit.
From product catalogs or examination of units, select a heater whose capacity is 3710 W or greater.

Unit Location
An individual heating/cooling unit is mounted somewhere in the building envelope, usually under a window if the sill is high enough. Otherwise, it is located in a part of the wall where its grillage can easily disperse the conditioned air into the heart of the room.

HOT WATER HEATING

Hot water heating design includes:

1. Boiler capacity
2. Boiler floor area
3. Boiler flue size and height, if not electrically operated
4. Heater unit size for each space

The *general recommendations* for a hot water heating system are:

1. Supply water design temperature is usually 170, 180, 190, or 200°F, with a 20° drop assumed through the zones. Higher temperatures are desired in systems having larger zones.
2. Water velocity should be between 1 and 4 gal/min. A higher velocity will create noise; a lower velocity will require larger piping.

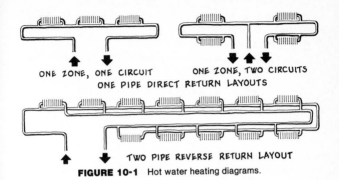

ONE ZONE, ONE CIRCUIT   ONE ZONE, TWO CIRCUITS
ONE PIPE DIRECT RETURN LAYOUTS

TWO PIPE REVERSE RETURN LAYOUT

FIGURE 10-1 Hot water heating diagrams.

If a zone has more than three units in any circuit, a two-pipe reverse return piping layout (Fig. 10-1) should be used.

3. Straight runs of longer than 30 ft should be avoided because of problems that arise from thermal expansion.

4. All piping runs should be pitched at a minimum ⅛-in/ft, and vents should be installed at the peaks so that trapped air in any part of the system can be released.

5. Valves should be installed at all low points in the piping layouts for draining the system during periods of shutdown.

6. Each zone should be valved so that each may be shut down without shutting down the whole system.

7. All supply and return piping should be insulated.

Boiler Capacity

The design heat load for a small industrial building near Boston is 271,000 BTU/hr. How big should the boiler for this building's hot water heating system be?

$$C_h = 1.15H + 0.00004HE$$

C_h = heating capacity of boiler (use net ratings as listed in manufacturer's catalog), ? BTU/hr

H = total design heat load, 271,000 BTU/hr

E = elevation above sea level (E for Boston is near zero), E = zero if elevation is less than 2000 ft above sea level

$$C_h = 1.15 \times 271{,}000 = 312{,}000 \text{ BTU/hr}$$

From product catalogs, select any water boiler whose capacity exceeds this amount.

Boiler Floor Area

What is the desired floor area for a water boiler whose base is 32 in wide and 34 in long?

$$A = \left(\frac{L}{12} + 5\right)\left(\frac{W}{12} + 5\right)$$

A = floor area for boiler and surrounding access, ? ft^2

L = length of boiler base, 34 in

W = width of boiler base, 32 in

$$A = \left(\frac{34}{12} + 5\right)\left(\frac{32}{12} + 5\right) = 60.0 \text{ ft}^2 \text{ minimum area}$$

NOTE: This area should be approximately square and does not include any chimney base.

Boiler Flue Size and Height

What is the chimney size for an H. B. Smith G300-7W boiler?

Somewhere in the H. B. Smith product catalog you can find this information. On the "Dimensions and Ratings" page, beside "G300-7" and under "chimney size," is the number "8x12x20." This means the flue should be 8 x 12 in in size and at least 20 ft tall.

Heater Unit Size for Each Space

The manager's office in a small industrial building has a design heating load of 5530 BTU/hr. The room, located at a corner of the building, has two windows on a long side and one on the end. You plan to install a water baseboard heater under each window. If the selected units are rated at 640 BTU/hr per lin ft at a 190°F operating temperature and 20° drop, what are their lengths?

$$H = LR$$

H = design heat load of the space, 5530 BTU/hr
L = total length of heating units installed in the space, **?** lin ft
R = rating of selected heating unit (as listed in product catalog), 640 BTU/hr per ft

$$5530 = 640L$$
$$L = 8.64 \text{ ft total length} \rightarrow 9 \text{ ft}$$

Use 3 ft under each window.

AIR HEATING

Air heating system design includes:

1. Furnace capacity
2. Furnace floor area
3. Furnace flue size and height, if not electrically operated
4. Duct sizing
5. Register locations and patterns

The *general recommendations* for an air heating system are:

1. Discharge air should be between 120° and 145°F. Higher temperatures result in high air velocities that create noise; lower temperatures result in low air velocities that require larger and more costly ducting.

2. In well-made systems, flexible connections that reduce sound vibration exist between the furnace and main ducts.
3. All ducting should be wrapped with thermal insulation that also reduces sound transmission.
4. Ducts are most efficient if round or nearly square. Above 6 in, most square or rectangular ducts are made in multiples of 2 in.

Furnace Capacity

A residence in Denver, Colorado, has a design heating load of 61,900 BTU/hr. If the house has an air heating system, what is the minimum capacity of the furnace?

$$H_c = 1.15H + 0.00004HE$$

H_c = heating capacity of furnace (use net ratings as listed in product catalogs), ? BTU/hr
H = total design heat load, 61,900 BTU/hr
E = elevation above sea level, in Denver E = 5280 ft

$$H_c = 1.15 \times 61,900 + 0.00004 \times 61,900 \times 5280$$
$$= 84,300 \text{ BTU/hr minimum capacity}$$

Furnace Net Area

Follow the procedure given in "Boiler Floor Area" under Hot Water Heating, page 412.

Furnace Flue Size and Height

Follow the procedure given in "Boiler Flue Size and Height" under Hot Water Heater, page 412.

Duct Sizing

Example 1 The capacity of a residential furnace is 89,000 BTU/ hr. If the furnace heats the supply air to 130°F, the air flows out of two

TABLE 10-3 DUCT VELOCITY FACTORS*

Architecture	Main ducts	Branch ducts	Outlet ducts
Residences	8	5	3.5
Theatres, assembly areas	8	6	4
Apartments, hotel and hospital bedrooms	10	7	5
Private offices, conference rooms, libraries, schools	12	8	6
General offices, banks, fine restaurants and stores	15	9.5	7
Average retail, cafeterias	18	12	9
Industrial, recreational, service areas	22	18	12

*For low-velocity systems.

main ducts of equal size, and the indoor temperature is 68°F, how large are the ducts?

$$C_h = 0.65 AD(T_h - T_i)$$

C_h = heat capacity of the furnace, 89,000 BTU/hr

A = cross-sectional area of ducting (outside dimensions minus insulation), ? in^2

D = duct velocity factor (as listed in Table 10-3 for main duct in residence), 8

T_h = temperature of heated air, 130°F

T_i = temperature of indoor air, 68°F

$$89,000 = 0.65 \times A \times 8(130 - 68)$$
$$A = 276 \text{ in}^2 \text{ for both main ducts}$$

One duct = $276/2$ = 138 in^2 → 12 x 12 in, 10 x 14 in, or 14-in diameter.

NOTE: For long (more than 40 ft) or unusual ducting, sizing should consider air friction losses and aspect ratio, as described in ''Supply Duct Size'' under Air Heating/Cooling: HVAC, pages 423-425.

Example 2 A bedroom in the above residence has a design heat load of 6600 BTU/hr. If the air enters this space through two ducts of equal size, what are their sizes? Remember, duct temperature = 130°F and room temperature = 68°F.

$$H = 0.65AD(T_h - T_i)$$

H = design heat load of the space, 6600 BTU/hr

A = cross-sectional area of the ducting, ? in^2

D = duct velocity factor (as listed in Table 10-3 for outlet duct in residence), 3.5

T_h = temperature of heated air, 130°F

T_i = temperature of room air, 68°F

$$6600 = 0.65 \times A \times 3.5(130 - 68)$$
$$A = 46.8 \text{ in}^2 \text{ for both ducts}$$

One duct = 46.8/2 = 23.4 in^2 → 4 x 6 in, 3 x 8 in, or 6-in diameter

Register Design

In air heating, wall-mounted registers should be close to floors. Floor-mounted registers should be near walls, preferably in outer walls and under windows.

Grillage should direct incoming air across nearby wall or floor surfaces or in directions that avoid drafts and give the space maximum air circulation.

For more on register design, see "Register Design" under Air Heating/Cooling: HVAC, pages 429-431.

AIR COOLING

Air cooling system design includes:

1. Sizing the cooling unit
2. Sizing the ducts, if any extend from the unit
3. Designing the registers

The *general recommendations* for an air cooling system are:

1. Air-conditioning (AC) units are usually factory-made compact packages that are installed in the surface of the building envelope, generally in a window, wall, or flat roof. Usually the cooling air flows directly from the machine into an adjacent large open space. Sometimes, a small length of ducting carries part of the cooling air into nearby smaller enclosed spaces.
2. A good indoor design temperature for air conditioning is 78°F at 50 percent relative humidity.
3. Most AC units handle about 400 ft³/min of cooling air per ton of refrigeration.
4. All AC units are equipped with filter boxes.
5. Where AC ducts pass through unconditioned spaces, they must be insulated.

Unit Sizing

Example 1 The design cooling load of the living-den-dining area in a residence near Dallas is 50,800 BTU/hr. What is the minimum capacity of a unitary air conditioner mounted in one of the living room windows?

$$C_c = 1.15C + 0.00004CE$$

C_c = minimum cooling capacity of the AC unit, ? BTU/hr
C = design cooling load, 50,800 BTU/hr
E = elevation above sea level (elevation of Dallas is 500 ft), zero if less than 2000 ft

$$C_c = 1.15 \times 50,800 = 58,400 \text{ BTU/hr}$$

As AC unit capacities are rated in tons, convert BTU/hr into tons:

$$1 \text{ ton} = 12{,}000 \text{ BTU/hr}$$

$$\frac{\text{BTU/hr}}{12{,}000} = T$$

$$\frac{58{,}400}{12{,}000} = 4.87 \text{ tons}$$

4.87 tons is the minimum capacity of the AC unit to be installed in one of the living room windows. However, the largest window AC unit made is 2.5 tons. Therefore, install two 2.5 ton units in two windows of the space.

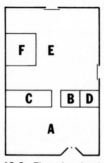

FIGURE 10-2 Floor plan showing rooms.

Example 2 A one-floor building housing an auto dealership near Atlanta is to be air-conditioned. Figure 10-2 is a plan of the structure. A list of the structure's spaces and their design cooling loads is given below.

Space	AC unit cooling load, tons
A: Showroom	10.38
B: Owner's office	0.53
C: Sales and secretarial	1.09

D:	Service manager's office	0.39
E:	Garage	14.51
F:	Parts department	1.44

Two rooftop cooling units shall be mounted above the showroom and four above the garage. Considering this arrangement, what are the desired capacities of the units and how should all the spaces be cooled?

STRATEGY Try to find a unit whose tonnage will supply all of a larger space (showroom or garage) and one or more of the smaller spaces.

STEP 1.   Check the showroom capacities.

$$\frac{\text{Tons load}}{\text{Number of units}} = \frac{10.38}{2} = 5.19 \text{ tons per unit}$$

Two 5-ton units would work here, with no cooling remaining for other spaces.

STEP 2.   Assume that the parts department will be cooled by the garage units, and check the total capacities of these areas.

$$\frac{\text{Tons load}}{\text{Number of units}} = \frac{14.51 + 1.44}{4} = \frac{15.95}{4} = 3.99 \text{ tons each}$$

Possibly, if 5-ton units were also used for these areas, enough cooling would remain to heat the three smaller spaces.

STEP 3.   Add the capacities of the remaining spaces (B, C, and D) to the garage area load and divide by four units.

$$\frac{\text{Tons load}}{\text{Number of units}} = \frac{0.53 + 1.09 + 0.39 + 15.95}{4}$$

$$= \frac{17.96}{4} = 4.49 \text{ tons each}$$

As 4.49 tons is less than 5, four 5-ton units would work nicely for all the spaces B through F. Short ducts would run from the garage units into the smaller spaces nearby, as shown in Fig. 10-3.

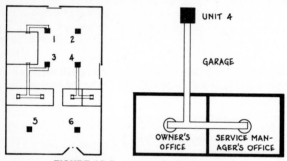

FIGURE 10-3   Large and small floor plans.

Duct Sizing

Example 1   One-sixth of the total area of a one-floor building is cooled by a roof-mounted 5-ton AC unit, as shown on the right of Fig. 10-3. If most of the cooling from this unit pours into the large area and the rest flows into the two smaller spaces whose loads are 0.39 and 0.53 tons, what is the size of the main duct leading from the AC unit into the smaller spaces?

$$803T = AD$$

T = tonnage capacity of cooled spaces, $0.39 + 0.53 = 0.92$ tons
A = cross-sectional area of main duct, ? in^2
D = duct velocity factor from Table 10-3, D for main duct passing through industrial area = 22; D for main duct passing through private office = 12; use smaller value, $D = 12$

$$803 \times 0.92 = A \times 12$$
$$A = 61.6 \text{ in}^2$$

Ducts should be 8 x 8 in, 6 x 12 in, or 9-in diameter.

NOTE: For long (more than 40 ft) or unusual ducting, sizing should consider air friction losses and aspect ratio, as described in "Supply Duct Size" under Air Heating/Cooling: HVAC, pages 423-425.

Example 2   What is the minimum size of an outlet duct entering a 13 x 15 ft office that has a design heat load of 5400 BTU/hr?

$$C = 15AD$$

C = design cooling load for space, 5400 BTU/hr
A = cross-sectional area of main duct, ? in^2
D = duct velocity factor (as listed in Table 10-3, D for outlet duct in private office), 6

$$5400 = 15A \times 6$$
$$A = 60 \text{ in}^2$$

The duct should be 8 x 8 in, 6 x 10 in, or 9-in diameter.

Register Design
Cooling registers should be located in ceilings or at least 6.5 ft high on walls. Wall registers should be spaced about 10 ft apart to prevent dead spots. Round ceiling outlets should be located near the center of the zone.

Register grillage should prevent drafts from forming in the lower 6 ft of space. This is best done by directing air across nearby wall or ceiling surfaces.

For more on register design, see "Register Design" under Air Heating/Cooling: HVAC, pages 429-431.

AIR HEATING/COOLING: HVAC

HVAC design includes:

1. Sizing the HVAC unit
2. Establishing fresh air requirements
3. Sizing supply ducting
4. Sizing return ducting
5. Selecting desirable duct layouts and fittings
6. Designing and locating registers

The *general recommendations* for an HVAC design are:

1. HVAC units are centralized systems for large buildings.
2. Recommendations for air heating and air cooling systems hold for HVAC systems.
3. Desirable operating temperatures for air HVAC systems are:

Heating air	120 to 145°F
Heating indoor temperature	67 to 72°F at 50% R.H.
Cooling air	50 to 60°F
Cooling indoor temperature	75 to 80°F at 50% R.H.

Unit Sizing

A corporate office building near St. Louis has a design heating load calculated at 1,192,000 BTU/hr and a design cooling load of 742,000 BTU/hr. If the structure has a centrally located variable air volume HVAC system, what are its design heating and cooling capacities?

Heating capacity:

$$C_h = 1.15H + 0.00004HE$$

Cooling capacity:

$$C_c = 1.15C + 0.00004CE$$

C_h = minimum heating capacity of the HVAC system, ? BTU/hr
C_c = minimum cooling capacity of the HVAC unit, ? BTU/hr
H = design heating load, 1,192,000 BTU/hr

C = design cooling load, 742,000 BTU/hr
E = elevation above sea level (elevation of St. Louis is 446 ft), zero if less than 2000 ft

$$C_h = 1.15 \times 1,192,000 = 1,370,000 \text{ BTU/hr}$$
$$C_c = 1.15 \times 742,000 = 853,000 \text{ BTU/hr}$$

Fresh Air Requirements
See the section on Ventilation, page 439.

Supply Duct Size
A large office building near St. Louis contains a low-velocity variable-air-volume HVAC system. The equipment's heating capacity is 1,370,000 BTU/hr and its cooling capacity is 853,000 BTU/hr. If the conditioned air leaves the central system via two main ducts, one of which carries 53.7 percent of the load, what size are the two ducts? Each duct is 65 ft long and cannot be deeper than 24 in; heating condition temperatures are supply air = 125°F and indoor air = 68°F.

STEP 1. Find the cross-sectional area of the two ducts from the two formulas below. Use the larger A.

Heating:

$$C_h = 0.65AD(T_h - T_i)$$

Cooling:

$$C_c = 15AD$$

C_h = heating capacity of HVAC system, 1,370,000 BTU/hr
C_c = cooling capacity of HVAC system, 853,000 BTU/hr
A = area of ducting (both main ducts), ? in^2
D = duct velocity factor (from Table 10-3, D for main duct in general office area), 15
T_h = heating air temperature, 125°F
T_i = indoor temperature, 68°F

Heating:

$$1,370,000 = 0.65A \times 15(125 - 68)$$
$$A = 2470 \text{ in}^2$$

Cooling:

$$853,000 = 15A \times 15$$
$$A = 3790 \text{ in}^2$$

Use the larger A: 3790 in^2
   Larger duct area:

$$53.7\% \text{ of } 3790 = 2040 \text{ in}^2$$

Smaller duct area:

$$3790 - 2040 = 1750 \text{ in}^2$$

STEP 2.   Find the duct friction factor from the formula below.

$$F = \frac{0.09L^{0.8}}{D^{0.15}}$$

F = duct friction factor (for each duct), ? in
L = length of each duct, 65 ft
D = duct velocity factor (same as in step 1), 15

$$F = \frac{0.09(65)^{0.8}}{(15)^{0.15}} = 1.69 \text{ in}$$

STEP 3.   Find the net area of the duct from the formula below.

$$A_n = (\sqrt{A} + F)^2$$

A_n = net area of each duct, ? in^2
A = cross-sectional area of each duct as found in step 1 above (from this point on, only calculations for the larger duct will be performed); area of larger duct = 2040 in^2

F = duct friction factor as found in step 2, 1.69 in

$$A_n = (\sqrt{2040} + 1.69)^2 = 2200 \text{ in}^2$$

STEP 4.   Give the duct its final dimensions according to the formulas below.

If the duct is round:

Formula A: $d = 1.13 A_n$

If the duct is rectangular and its aspect ratio (wider side length/narrower side length) = 2.5 or less:

Formula B: $A_n = NW$

If the duct is rectangular and its aspect ratio exceeds 2.5, increase its wider dimension by:

Formula C: $W = \dfrac{A_n}{N}\left[1 + 0.072\left(\dfrac{A_n}{N^2} - 1\right)\right]$

d = diameter of the duct, if round; duct is rectangular
A_n = net cross-sectional area of the duct, from step 3 above, 2200 in^2
N = narrower dimension of duct, if rectangular, 24 in maximum
W = wider dimension of duct, if rectangular, ? in

Aspect ratio:

$$\frac{W}{N} = \frac{WN}{NN} = \frac{A_n}{N^2} = \frac{2200}{24^2} = 3.8$$

Use formula C:

$$W = \frac{2200}{24}\left[1 + 0.072\left(\frac{2200}{24^2} - 1\right)\right]$$
$$= 110 \text{ in}$$

Final size of larger duct is 24 x 110 in.

Return Duct Sizing

In an enclosed, centralized air system, return airflow should be 80 percent of supply airflow in every space. Thus, the return outlets in a room should be 80 percent of the area of the supply outlets, return branch duct area should be 80 percent of supply branch duct area, and return main duct area should be 80 percent of supply main duct area.

For duct sizing of areas having special ventilation requirements see "Exhaust Duct Sizing" under Ventilation, pages 442-443.

Duct Layouts and Fittings

Elbows   (see Fig. 10-4)
90° rectangular elbow, for rectangular ducting

90° smooth elbow, for round ducting

Vaned square elbow, used in rectangular ducting where space prevents use of round elbows

45° smooth elbow, for round ducting

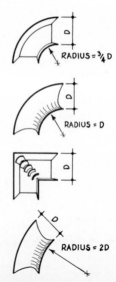

FIGURE 10-4   Duct elbows.

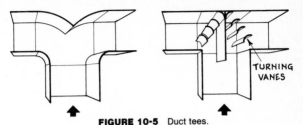

FIGURE 10-5 Duct tees.

Tees Used for rectangular ducting (Fig. 10-5).

Takeoffs These are duct intersections that contain a thermostat-controlled damper. They are needed when two spaces served by a common branch duct have differing heating and cooling loads. The illustration on the title page of this chapter shows several types of takeoffs and a heating/cooling condition that demonstrates the use of one. In winter, space A has a slightly greater heating load than space B, but in summer the cooling load in space B is three times that of space A. The takeoff always divides the incoming air according to the proportions needed by each space.

Mixing boxes Used for mixing the proper proportions of hot and cold air to create the desired temperature of the air flowing into the space (Fig. 10-6).

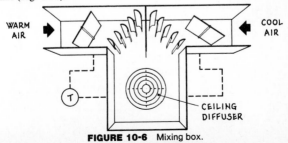

FIGURE 10-6 Mixing box.

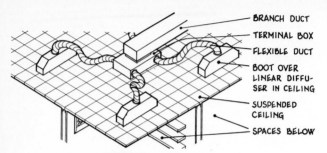

FIGURE 10-7   Flexible ducts.

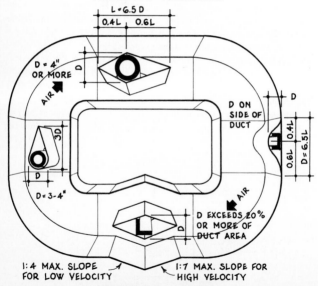

FIGURE 10-8   Easements.

Flexible ducts Used for versatile and economical dispersing of conditioned air to interior spaces (Fig. 10-7).

Easements Used for streamlining obstructions passing through ducts (Fig. 10-8).

Transformations Used for changing the shape of a duct when passing exterior obstructions (Fig. 10-9).

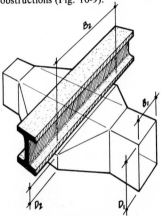

CROSS-SECTION AREAS...
$B_2 D_2 = 0.8 B_1 D_1$ MIN.

FIGURE 10-9 Transformation.

Register Design

For heating and cooling systems, the best register locations are high on walls or in ceilings. Grillage should direct incoming air across nearby wall or floor surfaces or in directions that minimize drafts yet provide maximum air circulation. Many grilles should have adjustable vanes.

Indoor air velocities should be between 15 and 50 ft/min, with 25 ft/min as ideal. Below 15 ft/min air feels stagnant; above 50 ft/min, light papers blow off a desk.

The throw of a wall register should be 0.75 the distance between its grill and the other side of the room, if no registers exist on the opposite side. Ideal throw at this point is 50 ft/min 6.5 ft above the floor. Register throws are usually listed in manufacturer's catalogs.

Grillage should have a spread that disperses the air usefully into the space. Three desirable vane angles are shown in Fig. 10-10.

The overall outside dimension of a register is usually 1.5 in longer and wider than the duct it fits into.

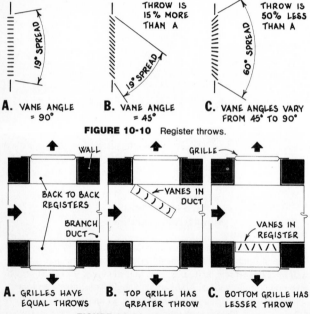

A. VANE ANGLE = 90°

B. VANE ANGLE = 45°

C. VANE ANGLES VARY FROM 45° TO 90°

FIGURE 10-10 Register throws.

A. GRILLES HAVE EQUAL THROWS

B. TOP GRILLE HAS GREATER THROW

C. BOTTOM GRILLE HAS LESSER THROW

FIGURE 10-11 Back-to-back grilles.

One particular design problem exists with back-to-back grilles. With registers in this position, either the throws of both must be equal or vanes must be positioned in the ducting to control air flow (Fig. 10-11).

Registers come in a variety of sizes and shapes. A sampling is shown in Fig. 10-12.

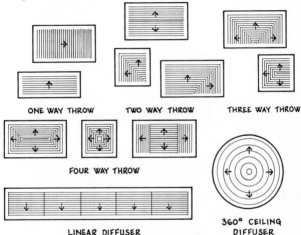

ONE WAY THROW   TWO WAY THROW   THREE WAY THROW

FOUR WAY THROW

LINEAR DIFFUSER

360° CEILING DIFFUSER

FIGURE 10-12 Register sizes and shapes.

AIR-WATER HEATING AND COOLING

Because air-water HVAC systems are complicated and require advanced engineering knowledge to design, their mathematics are too involved to be dealt with here. Instead, this section describes the nature of this system's components and the facts behind its operation.

Figure 10-13 is a schematic of an air-water HVAC system. A brief description of each component follows.

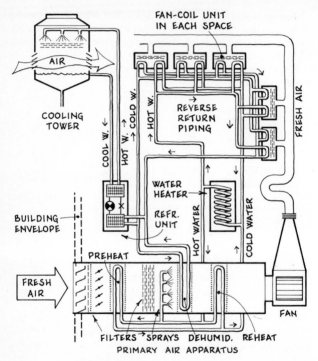

FIGURE 10-13 Air-water HVAC system.

Primary air apparatus This machinery resembles a large, long, rectangular tube with one end mounted in the building envelope. Here outdoor air enters through a louver at a face velocity of 500 to 800 ft/min and then passes through a filtering screen, dampers that control its volume, preheaters that in winter prevent freezing air from entering the dehumidifier, high-efficiency filters that remove dirt particles and entrained dust, sprays that add humidity in winter, dehumidifiers that remove water vapor in summer, a reheater that heats the air to about 85°F, and finally the primary fan that drives the air through the ducting.

The primary air apparatus is sized according to the building's fresh air requirements.

Refrigerating unit This unit makes chilled water for the fan-coil units in the occupied spaces and the dehumidifier coils in the primary air apparatus. Its evaporator cools the water to about 48°F, and the heat built up in the condenser is carried by water piping to the cooling tower.

The refrigeration load is determined by the maximum sum of the individual room peak loads (not the design peak load).

Cooling tower This dissipates the heat carried away by water from the condenser in the refrigeration unit. The warm water enters the top of the tower, passes through nozzles, falls as spray, is cooled by air flowing through the middle of the tower, collects in a basin at the bottom, and then flows back to the refrigeration unit. The tower, usually built on the roof, should stand at least 100 ft away from any chimneys or other sources of heat or contaminated air, and its air intakes should allow air to enter quickly and leave rapidly without recirculating. The tower should also have a city water hose bibb.

In below-freezing weather, cooling tower operation can create problems as a result of ice buildup in the basin, nozzle heads, and louvers.

The cooling tower is sized primarily by the condensing temperature in the refrigeration unit and the wet-bulb design temperature of the region. Higher condenser temperatures usually mean smaller cooling towers.

Water heater   This component heats the water pumped to the fan-coil units and sometimes supplies the heat for the preheater and reheater in the primary air apparatus. The temperature of water leaving the heater equals the highest temperature required in the fan-coil units plus 15 to 20 degrees.

The water heater is sized to handle the building heating load times 1.2 plus any water heat required to raise the temperature of the primary air.

Fan-coil units   This is typically a low, deep box located in each space, preferably under a window. In each unit, a small amount of primary air mixed with room air passes through a small radiator containing either hot or cold water, depending on the room air's temperature.

The size of each fan-coil unit depends on the ratio of required primary air to the total heating or cooling load of the room in which the unit is located.

Air ducting   Because of high air velocities (4000 to 5000 ft/min in the risers, about 3000 ft/min in the headers), rigid spiral tubing is used instead of sheet metal ducting; all fitting seams are welded to eliminate leakage.

Air ducting is sized according to the required volume of primary air and the air's velocity.

Water piping   The pipes carry the heat and cold from the water heater and refrigerating unit to the fan-coil units in each space. Each of these plumbing circuits requires a pump, and each layout on every floor should be reverse return. Proper allowances must be made for thermal expansion in all piping, and the total system requires an open expansion tank to permit air venting and water expansion caused by changes in temperature. All supply water piping should be wrapped in minimum 1-in insulation.

Water piping is sized according to the required heating and cooling loads and the water's velocity.

Example A developer plans to build an office building on a corner lot in the downtown area of a medium-sized city. The city building code describes setbacks and height limits for the property, as shown in Fig. 10-14.

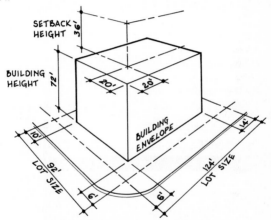

FIGURE 10-14 Building envelope perspective.

Conventional steel frame construction in the area requires about an 11.5 ft minimum floor-to-floor height, which means the developer could fit only six floors into the 72-ft envelope height limit. However, he believes there must be a way to squeeze seven floors into the envelope, with each level measuring about 8.5 ft from floor to ceiling and 18 to 20 in from the ceiling to the finished floor above.

What kind of structural and HVAC systems would work here?

STRATEGY   The structural system could be reinforced concrete flat plate with a 6-in floor slab, 8-in thick drop panels, and column-to-column spans 24 ft long both ways. The construction measures 18 in from bottom of ceiling to top of finished floor, and this includes a 9-in clearance for the water pipes and primary air duct of an air-water HVAC system. Around the edge of each floor, a short cantilever provides room for the system's water supply and return pipes, a supply air header, and fan-coil housings that provide individual climate control to each perimeter space.

A section through a cantilever is shown in Fig. 10-15.

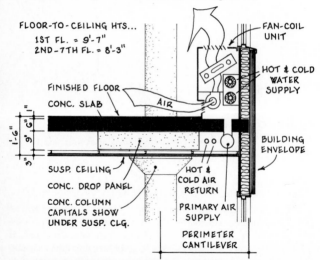

FIGURE 10-15   Section through fan-coil unit.

As for the other parts of the system, the cooling tower goes on the roof and the rest of the equipment is in the basement. With this arrangement, the perimeter apparatus takes up 2 or 3 percent of the gross floor area, the interior bundling of piping and ducting requires about 2 percent more, while electrical and plumbing shafts take up another 1 percent. Shaft dimensions include the surrounding walls. The basement is usually 3 to 5 percent of the gross building area, and this room's height is 13 to 18 ft. The cooling tower requires about 1 ft^2 of roof area per 400 ft^2 of gross building area, is from 15 to 40 ft high, and when full weighs 125 to 200 lb/ft^2.

HEAT PUMP

Heat pump design includes:

1. Selecting the capacity and number of units
2. Locating the components
3. Sizing the ducting
4. Locating the outlets and selecting the grilles

General recommendations for a heat pump design are:

1. Heat pump capacities range from about 15,000 to 50,000 BTU/ hr for heating and cooling loads. The larger load determines the unit's size. If the architecture requires more than two or three units, other systems are often more economical.
2. If winter temperatures in the region stay below 35°F for prolonged periods of time (a condition that usually exists in areas having more than 4000 heating degree-days), other systems are probably more economical. However, if the compressor coil can be immersed several feet deep in a large swimming pool or similar body of water, the system may be able to operate efficiently in much lower air temperatures.
3. If winter temperatures in the region go below about 35°F, heat pump operation must be assisted by a second heating system,

usually electric resistance fixtures. Below 15°F, the second heating system must be designed to supply the total heating load.

Maximum floor area per heat pump unit is about 1800 ft². Above this, ducting expense and sometimes less flexible climate control become limiting factors.

Capacity and Number of Units

A 2600-ft² residence near Nashville, Tennessee, is to be heated and cooled by heat pumps. If the house's design heating and cooling loads are 57,000 BTU/hr and 73,000 BTU/hr, how many heat pumps should the house have and what sizes should they be?

Size the units according to the higher capacity:

$$H_c = 1.15H \qquad C_c = 1.15C$$

H_c = heating capacity of the heat pump, ? BTU/hr
C_c = cooling capacity of the heat pump, ? BTU/hr
H = heating load of the architecture, 57,000 BTU/hr
C = cooling load of the architecture, 73,000 BTU/hr

$$C_c = 1.15 \times 73,000 = 84,000 \text{ BTU/hr}$$

The house needs two or three units.

$$\frac{84,000}{2} = 42,000 \text{ BTU/hr} \qquad \frac{84,000}{3} = 28,000 \text{ BTU/hr}$$

The house could have two units at about 42,000 BTU/hr cooling capacity each, or three units at about 28,000 BTU/hr cooling capacity each.

Locating the Components

A heat pump unit has two parts: the *compressor* and the *blower*. The compressor should be located outdoors, preferably close to the building and behind some planting or other shielding device. The blower belongs indoors, near the center of its zone so that any ducting is as short as possible.

Duct Sizing

Heat pump ducting is sized the same as air heating and cooling systems. The basic formulas, and the sections explaining their use, are listed below.

Heating:

$$H = 0.65AD(T_h - T_i)$$

(See "Duct Sizing" under Air Heating, pages 413-416.)

Cooling:

$$C = 15AD$$

(See "Duct Sizing" under Air Cooling, pages 420-421.)

Desirable duct layouts and fittings are described under "Duct Layouts and Fittings" in the section on Air Heating/Cooling: HVAC, pages 426-429.

Register Design

See the section on "Register Design" under Air Heating/Cooling: HVAC, pages 429-431.

VENTILATION

Ventilation design involves:

1. Establishing fresh air requirements
2. Sizing the exhaust fan
3. Sizing the exhaust ducting
4. Locating the exhaust system components

The *general recommendations* for ventilation are:

1. There are two kinds of ventilation: *fresh air* (which satisfies human fresh air requirements) and *exhaust* (which removes odors, smoke, fumes, and other undesirable air generated in a space). In fresh air ventilation, supply air comes from outdoors; in exhaust ventilation, removed air is emptied into the outdoors. Louvers for the former must be located far from louvers for the latter and in ways that prevent exhaust air from circulating into fresh air ducts.

2. In central HVAC systems, required fresh air is added to interior spaces and ordinary amounts of undesirable air are removed from spaces by the HVAC equipment. For unusual amounts of undesirable air occurring at periodic intervals (such as occasionally exists in a conference room filled with smokers), spaces must have separate equipment that draws the excess undesirable air into the outdoors.

Fresh Air Requirements

What are the ventilation requirements of an office conference room that can seat 24 people?

$$V = NR$$

V = total ventilation requirements of the space, ? ft³/min
N = number of people occupying the space, 24 people maximum
R = recommended ventilation requirements in cubic feet per minute (as listed in Table 10-4, R for conference room), 50 ft³/min per person

$$V = 24 \times 50 = 1200 \text{ ft}^3/\text{min}$$

Exhaust Fan Sizing

Example 1   The painting room in an adding machine factory is 19 ft long, 12 ft wide, and 10 ft high. How large should its exhaust fan be?

TABLE 10-4 RECOMMENDED VENTILATION REQUIREMENTS

Occupancy	Recommended requirement
	ft^3/min per person:
Inactive areas, sleeping, theatres with no smoking	7½
General office areas, retail, public with no smoking, libraries, classrooms	10
Apartments, barbershops, beauty parlors, drugstores, cafeterias, most factories	15
Hospital wards, laboratories, restaurants, light activity with slight odor generation	20
Private offices, bars, locker rooms, activities with moderate odor generation	25
Shipping rooms, hotel rooms, areas of active work	30
Conference rooms, meeting rooms with heavy smoking, gymnasiums, strenuous activity areas	50
Corridors	0.33 ft^3/min per ft^2 floor area
Garages	1.5 ft^3/min per ft^2 floor area
Commercial kitchens	4.0 ft^3/min per ft^2 floor area
Rest rooms	50 ft^3/min per water closet or urinal
Localized bad air	6–60 air changes per hour
Canopy hoods	60 ft^3/min per ft^2 of hood area

$$Q = 0.02NV$$

Q = fan air delivery rate, ? ft³/min
N = minimum number of air changes per hour required for the space
    (as listed in Table 10-4, N for localized bad air), 6 air changes
    per hour minimum; use 2.5 this amount to be "safe"
V = volume of space to be ventilated, $19 \times 12 \times 10 = 2280$ ft³

$$Q = 0.02 \times 6 \times 2.5 \times 2280$$
$$= 685 \text{ ft}^3/\text{min minimum}$$

Example 2 A men's room in an office building requires an exhaust
fan. If the room has seven toilets and five urinals, how big should the
exhaust fan be?

$$Q = 1.15VN$$

Q = fan air delivery rate, ? ft³/min
V = volume of air change required per unit (as listed in Table 10-4,
    V for a toilet or urinal), 50 ft³/min for each
N = number of units requiring ventilation, 7 toilets + 5 urinals = 12
    units

$$Q = 1.15 \times 50 \times 12$$
$$= 690 \text{ ft}^3/\text{min minimum}$$

Exhaust Duct Sizing

Example 2 A ventilation fan mounted in the center of the ceiling
in a men's room is rated at 720 ft³/min. If the exhaust air must pass
through 24 ft of ducting to a louver in the exterior wall, what size cross
section should the ducting have?

$$Q = 0.8DA$$

Q = fan air delivery rate, 720 ft³/min

D = duct velocity factor (as listed in Table 10-3, D for an outlet duct in a service area), 12

A = cross-sectional area of the exhaust duct, **?** in^2

$$720 = 0.8 \times 12A$$
$$A = 75 \text{ in}^2$$

The duct should be 8 x 10 in, 6 x 14 in, or 10-in diameter.

Example 2 When fully occupied, a conference room with 24 seats requires 1200 ft^3/min of fresh air. The central HVAC system continually supplies 240 ft^3/min of fresh air into the room. If an exhaust fan removes the excess smoky air when required, what size should it be?

Fan capacity = required ventilation − existing ventilation
= 1200 − 240 = 960 ft^3/min

Component Location

When an exhaust fan requires ducting, the farther the fan is located away from interior spaces, the less noisy it will be.

When removing heated air, the inner opening of the exhaust stream should be located as close as possible to the source of heat.

The outer opening of the exhaust stream must always be protected from entry by insects, debris, and undesirable weather. With rain or snow, in some locations updrafts against the surface of the building must be considered.

FILTRATION

In mechanical heating and cooling equipment, filters remove odors, chemicals, and particles suspended in fresh air from the outdoors and return air from the indoors. Filters are rated in terms of useful life span, resistance to airflow, and contaminant arrestance (efficiency). Although efficiency is the primary criterion, high efficiency usually means high airflow resistance, low dust holding capacity, and short life. Normal office use requires 30 to 85 percent efficiency.

In all mechanical equipment, filters should be accessible for service and replacement. In large units, inspection and service access should exist before and after the filter bank.

Pressure drop through the filter must be considered when calculating the fan velocity of the system.

Filters are selected according to the following procedure.

1. Determine the size, concentration, and character of the contaminants in the outdoor air and return air. This may be done by laboratory analysis, past experience, or general data.
2. Determine the size of the particles to be removed and efficiency of removal.
3. Select a filter that combines the greatest economy and efficiency under prevailing conditions of installation.

Below are descriptions of the most common filters used today and the positive and negative aspects of each.

Viscous impingement This is a relatively coarse and durable filtering medium that is periodically covered with grease or oil.

+ Good for removing pollens, dusts, ashes, mists, oil smokes, and visible particles. Especially suited for large particle contaminants of high concentration. Enjoys long life and great capacities; requires low service. Good prefilter for other type filters.
− Not suitable for linty atmospheres, viruses, bacteria, tobacco smoke, toxics, or lung-damaging particles.

Cleanable media Removable panels that can be readily cleaned and oiled. Used for airflow velocities of about 300 ft/min and is 65 to 80 percent efficient.

+   Offers maintenance economies if staff and cleaning space are available.
−   Not good where above conditions do not exist.

Automatic Overlapping filter panels attached to a chain that moves them across the airstream and through an oil bath. This type is for airflow velocities of about 500 ft/min and is 80 to 90 percent efficient.

+   Airflow resistance remains fairly constant; filter is constantly clean.
−   Costly, bulky, requires electricity.

Dry media This type has a permanent frame containing a dry replaceable filtering medium of cellulose, glass fibers, treated paper, cotton batting, wool felt, charcoal, or synthetics.

+   Good for removing microscopic particles of light concentration. These filters remove much of what viscous impingement types do not remove.
−   Small holding capacities, short life, high servicing costs.

Throwaway Packets of glass fiber or other medium 2 to 6 in deep. These are good for 400 to 500 ft/min airflow velocities and are 3 to 20 percent efficient.

+   Good for low dust loadings and prefiltering in large systems. Easily replaceable, low unit costs.
−   Inefficient, do not last long.

Roll-type disposable   Moving filter rolls that pass through the airstream and collect on a takeup spool as it becomes clogged. This method is good for medium airflow velocities and is 20 to 50 percent efficient.

+ Low replacement factor, constantly clean, constant airflow resistance, economical prefilter.
− Rather low efficiency.

Activated charcoal   Replaceable cartridges with perforated faces and interior of activated charcoal. They are useful at low airflow velocities and are highly efficient for specific toxic removals.

+ Good for specific odor and toxic removal.
− Greatly reduces airflow rates.

High-efficiency particulate (HEPA)   Deep mats or blankets with high ratio of filter media to duct area. These are useful at low airflow velocities (around 10 ft/min and are up to 99.97 percent efficient.

Electronic   This type removes airstream particles by electrically charging them and then attracting them to a medium of opposite charge located within the airstream.

+ Very efficient at microscopic particle range, induces very low pressure drops within the airstream.
− High initial and operating costs, air at higher than 70 percent relative humidity may adversely affect operation, prefilters required in linty atmospheres, not recommended for white rooms unless after-filters are used.

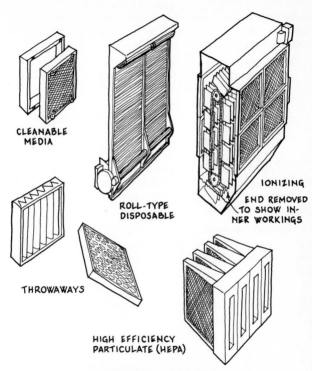

CLEANABLE
MEDIA

ROLL-TYPE
DISPOSABLE

IONIZING

END REMOVED
TO SHOW IN-
NER WORKINGS

THROWAWAYS

HIGH EFFICIENCY
PARTICULATE (HEPA)

FIGURE 10-16 HVAC filters.

Ionizing Removes contaminating particles by passing the air-stream through groups of metal plates charged at about 12,000 V. The plates are rotated by a chain that immerses them in an oil bath. This method is useful at high airflow velocities and is 85 to 90 percent efficient.

+   Good for systems that have large air deliveries containing high concentrations of fine dust particles. Suitable where equipment is relatively inaccessible or where service is infrequent. Has very low resistance to airflow.
−   Useless during periods of electric power outage. Afterfilters usually required to catch large particles that flake off the plates. Requires extra space for transformer and rectifier.

Charged media   Panel filters containing a dry medium electrostatically charged at 12,000 to 13,000 V. Power consumption is about 8 W per 1000 ft^3/min. Useful at medium to high velocities and is about 60 percent efficient (Fig. 10-16).

+   Combines qualities of viscous impingement and ionizing type filters, partially effective during power outages.
−   Requires extra space for transformer and rectifier.

INDEX

About the Author

Robert Brown Butler is an architect with twelve years experience as a building contractor and seven years experience in commercial architectural offices. He has also taught solar energy and architectural drafting, designed and built solar houses, won two Federal HUD grants for solar architecture, and is the author of *The Ecological House*.

Metric Conversion Factors *(Continued.)*

Conversion of U.S. Customary Units to Metric Units

Relationships

1 kilometer (km) = 1000 meters (m)
1 meter = 100 centimeters (cm)
1 centimeter = 10 millimeters (mm)
1 square kilometer Km²) = 100 hectares (ha)
1 hectare = 10,000 square meters (m²)
1 liter (L) = 1000 cubic centimeters (cm³) = 1/1000 cubic meter (m³)
1 newton (N) = 1 kilogram-meter per second (kg • m/s) = 100,000 dynes (dyn)
1 joule (J) = 1 newton-meter (N • m) = 10,000,000 ergs

*Conversion factors are taken to *four* significant figures.

†Measurement of land areas and water surfaces in metric units may also be expressed
 in the unit hectare (ha), which equals 10,000 m² (exactly). Therefore, 1 acre is equal to
 0.4047 ha.

‡The *board foot* measures a theoretical volume of 12 in x 12 in x 1 in for wood products.

§The *liter* (L) is a special name for the cubic decimeter (dm³) and equals 0.001 m³
 (exactly). It is used for liquid and dry measurement of volume and may also be used
 with the prefixes milli and micro (1 mL = 1000 mm³; 1 μL = 1 mm³).